The Political Economy of Nuclear Energy

Dipak Basu · Victoria W. Miroshnik

The Political Economy of Nuclear Energy

Prospects and Retrospect

palgrave
macmillan

Dipak Basu
Nagasaki University
Nagasaki, Japan

Victoria W. Miroshnik
Reitaku University
Chiba, Japan

ISBN 978-3-030-27028-5 ISBN 978-3-030-27029-2 (eBook)
https://doi.org/10.1007/978-3-030-27029-2

This Palgrave Macmillan imprint is published by the registered company Springer Nature Switzerland AG
The registered company address is: Gewerbestrasse 11, 6330 Cham, Switzerland

This book is dedicated to our colleague and our dear friend, Angelica Mari-Isabella Boss, the student of the Gymnasium 1811 Moscow, Russia, who, in fact, was an author of the idea of this book and who was an active participant in the research process, namely, she was responsible for collection of the research data, as well as translating numerous materials from Russian language into English. She is a real scientist despite her young age. Her bright insights and strong enthusiasm made this research project a success.

Contents

1 Introduction 1

Part I Political Economy of Nuclear Power

2 Advantages of Nuclear Power 7

3 Geopolitical Aspects of Nuclear Power 23

4 Economic Aspects of Nuclear Power 35

5 Military Aspects of Nuclear Energy 43

Part II Retrospect of Nuclear Power Development

6 History of the Development of Nuclear Power 61

7 Global Nuclear Policy Development 95

8 Global Nuclear Investment Environment 113

Part III Prospect of Nuclear Power Development

9 Germany as Nuclear Power 131

10 The United States as Nuclear Power 159

11 Russia as a Nuclear Power 181

12 Japan as Nuclear Power 197

13 India as a Nuclear Power 215

14 China as Nuclear Power 229

15 Future of Nuclear Power Development 247

Conclusion 261

Appendix 267

References 269

Index 273

Abbreviations and Explanations

ABM	Anti-ballistic missile defense systems
ABWR	Advanced boiling water reactor
ASEAN	The Association of Southeast Asian Nations
ASNAC	Strategic Nuclear Forces
BR-1	Breeder (fast-neutron) reactor
BREST	Reactors on a lead coolant
BWR-5 type reactor	Boiling water reactor
C^4ISTAR	Command, Control, Communications, Computers, Intelligence, Surveillance and Reconnaissance
CAEA	China Atomic Energy Authority World Nuclear Association
CEN	Combat Management for the Unified Space System of Russia
CIS	The Commonwealth of Independent States (CIS); 10 post-Soviet republics in Eurasia formed following the dissolution of the Soviet Union.
CNEA	China Nuclear Energy Association
CO_2	Carbon dioxide
DUHF	Depleted uranium hexafluoride

Euratom	The European Atomic Energy Community (EAEC or Euratom) is an international organisation established by the Euratom Treaty on March 25, 1957 with the original purpose of creating a specialist market for nuclear power in Europe, by developing nuclear energy and distributing it to its member states while selling the surplus to non-member states.
FBR	Fast breeder reactors
HEU	Highly enriched uranium
IAEA	International Atomic Energy Agency
ICBM	Intercontinental Ballistic missiles
ICREI	International Commission on Radiation Units and Measurements
IEA	International Energy Agency
IER	Institute for Energy Research of Washington, DC
INPRO	Innovative nuclear power project
LEU	Low-enriched uranium
LMFBR	Liquid Metal Fast Breeder Reactor
Magnox	160 MW gas cooled reactor
MW	Megawatt
NATO	North Atlantic Treaty Organization
NCMC	National Center for Defense Management of Russia
NDRC	National Energy Development Research of China
NNW	Non-strategic nuclear weapons
NPP	Nuclear Power Plants
NPT	Treaty on the Non-Proliferation of Nuclear Weapons
NSG	Nuclear Suppliers Group
NSNF	Strategic Nuclear Forces
NW	Nuclear weapons
OPEC	Organization of Petroleum Exporting Countries
Pu239	Plutonium-239
PWR	Reactor unit with pressurized water
RBMK	(Russian: Реактор Большой Мощности Канальный, РБМК; *Reaktor Bolshoy Moshchnosti Kanalnyy*, “High Power Channel-type Reactor”) is a class of graphite-moderated nuclear power reactor designed and built by the Soviet Union

REGNUM	Russian non-governmental, nationwide online news service disseminating news from Russia and abroad from its own correspondents, affiliate agencies and partners
RES	Renewable energy sources
ROSATOM	Russian state corporation
SALT-1 (1972)	Strategic Arms Limitation Treaty-1 in 1972
SALT-2 (1979)	Strategic Arms Limitation Treaty-2 in 1979
SMF	Strategic Missile Forces
SNF	Strategic nuclear force
SNS	Strategic offensive forces
SSB and SSBN	*SS* denotes submarine (or submersible ship), the *B* denotes ballistic missile, and the *N* denotes that the submarine is nuclear powered.
SSBN	Ballistic missile submarine is a submarine capable of deploying submarine-launched ballistic missiles (SLBMs) with nuclear warheads.
START III Treaty	(Strategic Arms Reduction Treaty)
SVBR	Lead-bismuth eutectic Reactor
Th232	Thorium-232
TNT	Trinitrotoluene
TNW	Tactical nuclear weapons
TPP	Thermal power plants
TWh	Terawatt-hour (TWh), a measure of electrical energy, 10^{12} watt-hours
UN	United Nations Organization
UNEP	United Nations Environment Program
UNESCO	United Nations Educational, Scientific and Cultural Organization
Uranium-235	Fissile component,
Uranium-238	Natural uranium
USSR	Union of soviet Socialistic Republics.
WANO	World Association of Nuclear Operators
WMD	Weapons of mass destruction
WNA	World Nuclear Association

1
Introduction

There are two aspects of nuclear power development. The first aspect is the supply of massive amount of energy at a low cost to develop human civilization, the peaceful atom. The second aspect is the supply of, the military atom, massive amount of weapons to destroy the human civilization. This book is about understanding the relationship between developments of these two aspects of nuclear power in three perspectives: past, present, and future.

The development of nuclear energy to gain national competitive advantage started at the beginning of twentieth century and particularly during the two world wars: the World War I (WW1) and the World War II (WW2). The intensive research has been developing during the prewar years in Germany and Russia, and after revolution in 1917 in the Union of Soviet Socialistic Republics, the USSR. Later the research activities with the greatest scope and focus continued in the United States, particularly in view the fact many prominent European scientists, who migrated to the United States before and during war period, were involved in the research. Thus, it is important to note that the development of research on nuclear energy began with the creation of nuclear weapons, the impetus for the

D. Basu and V. W. Miroshnik, *The Political Economy of Nuclear Energy*,
https://doi.org/10.1007/978-3-030-27029-2_1

development of which was served by two world wars: the WW1 and the WW2.

The history of the expansion of nuclear power development focused on the development of the research on the creation of nuclear weapons of mass destruction. That work was carried out simultaneously worldwide by the scientists of the opposing sides—Nazi Germany under Hitler's leadership, as well as England's scientists led by Churchill, the United States under Roosevelt, and then Truman, and the USSR under Stalin. The research focusing on the first aspect of nuclear power development on peaceful atom to supply the massive amount of energy at a low cost to develop human civilization first in history started and successfully accomplished in the USSR under Stalin, with the creation on June 27, 1954, the fist in history of humankind, the Obninsk Nuclear Power Plant (NPP) to generate electricity from nuclear energy. Thus, based on historical facts it is obvious that Russia is the pioneer of development of peaceful atom, the first aspect of nuclear power development. The USA, the UK, Germany are the pioneers of the development of the military atom, the second aspect of nuclear power development.

The idea of creating nuclear weapons as source of national competitive advantage grew out in the West in developing the concept of "air war." As early as 1914, the famous English writer H. G. Wells in his science fiction novel "The World Set Free" (Wells 1914) predicted attack using atomic bombs. In 1918 the Italian military theorist General Giulio Douay argued that in the future war defeat to the enemy can be inflicted by strategic bombing called "carpet bombing" carried out from a safe distance for the attacking side aiming total destruction of enemy and enemy's infrastructure.

Using these theoretical suggestions, in WW2 the German air force destroyed in 1937 the entire city of Guernica in Spain and repeated that in Stalingrad, Russia during the devastating four days August 23–27 in 1942. However, during the WW2 the idea was to create a bomb that will allow winning the war at a lower cost. Thus, the development of the super bomb, the atomic bomb, has been intensified, particularly in the UK, the USA, and Germany.

Since then, the development of the air forces has been proceeding in two directions. First: the development of air defense from perspective to increase in the range and carrying capacity of bombers, which culminated in the creation in the 1940s of strategic aviation. Second: the increase in the power of aircraft warheads capable of hitting the enemy's strategic potential. The logical result of the second trend was the creation and testing of nuclear weapons in 1945 in the United States with dual purposes: the defeat of the Nazi Germany and Japan; the destruction of Russia, the USSR.

The United States successfully tested nuclear bomb in Japan by bombing two Japanese cities Hiroshima and Nagasaki in August 1945. However, the United States abandoned the idea according to the "Unthinkable" plan proposed by Churchill of launching a nuclear strike against the Soviet Russia immediately after the end of WW2 in 1945 when the retaliatory forces of the USSR were destroyed during the war, and again in the early 1950s, the most dangerous period of the Cold War. A decade later, Washington and Moscow again did not use the "window of vulnerability" to attack China, which was approaching the creation of nuclear weapons with the help of France.

It is believed, that in each case, the major world powers refused to use their military nuclear superiority, not because of fear of the nuclear retaliation, but because of the moral responsibility. For example, in his speech to parliament on March 1, 2019, President of Russia Vladimir Putin highlighted the fact that development of several new nuclear weapon systems is successfully completed with the intension to demonstrate to the world that nuclear forces are indispensable for Russia's security. The development of latest defense system, for example, the creation of S-500, which can protect Russia from hypersonic missiles and the corresponding weapons systems in the Unites States, shows the direction of the technology. However, according to top leader of Russia, the development of strategic nuclear weapons is Russia's only tool to inflict "predetermined damage" from potential enemy. However, Russian military doctrine never included a preventive nuclear strike or testing nuclear weapons on other countries. For the United States the situation is different, due to the presence of not only nuclear weapons, but also forward-based systems in Eurasia, nuclear allies in Europe and superiority in non-nuclear precision weapons.

The growth of world fuel and energy needs along with the substantial resource and environmental limitations of traditional energy necessitates the timely preparation of new energy technologies that can take on a substantial part of the energy requirements and stabilize the consumption of fossil fuels. Due to the accident in the Japanese nuclear power plant, the Fukushima Daiichi, confidence in the feasibility of using nuclear energy has been significantly shaken. However, despite the refusal of a number of countries of further exploitation of NPPs, a positive view in the development of world nuclear energy is noticeable, surprisingly in Japan.

Indeed, the forecasts of the International Atomic Energy Agency (IAEA), the global center for nuclear power cooperation, say that by 2030, up to 600 new NPPs will be built on the planet, while there are more than 436 today. The installed capacity of the global nuclear power industry currently is 390 gigawatts. If all this power were generated from coal and gas sources, then an additional 2 billion tons of carbon dioxide (CO_2) would be emitted annually into the atmosphere. However, according to estimates of the intergovernmental panel on climate change, all boreal forests (for example, the taiga forests located in Russia in the northern hemisphere of the planet) annually absorb only about 1 billion tons of CO_2, while all the forests of the planet can absorb about 2.5 billion tons of CO_2.

Reference

Wells, H. G. (1914). *The World Set Free*. London: Macmillan.

Part I

Political Economy of Nuclear Power

2
Advantages of Nuclear Power

Peaceful Atom

The large-scale development of nuclear energy has been driven by the need for electricity as a result of rapid industrialization, exhaustion of energy reserves, and severe political instability in those countries exporting oil and gas. In addition, traditional energy supplies originating from fossil fuels harm the environment in terms of greenhouse gas emissions.

Nuclear power, the result of energy released during the fission of heavy nuclei and the fusion of light nuclei (at least in principle), has changed the landscape that existed in the first half of the twenty-first century. Expansion of the field of application of nuclear power is primarily determined by its economic implication. Oil was expected at the beginning of the century to be depleted in 20–30 years. This did not take place; by contrast, the scale of oil production has increased enormously. The same story will be repeated with uranium.

Reducing the cost of nuclear electricity to make it sustainable will create new opportunities for electrical technology to be expanded. Since nuclear power plants of almost equal efficiency can be established anywhere in the world, electrical technology processes will be further accelerated as a result of new sources of raw materials being discovered. There will be several

D. Basu and V. W. Miroshnik, *The Political Economy of Nuclear Energy*,
https://doi.org/10.1007/978-3-030-27029-2_2

new technical processes that are not yet profitable for the production of electricity due to the high cost of production. For example, the prospect of reducing the cost of electricity will provide an opportunity to solve the daunting task of providing enough desalinated water for areas where water is scarce. The shortage of freshwater resources is rapidly growing and has affected both industrial and agricultural development. The first successful efforts to develop desalination technology on a large scale were taken in the former Soviet Union. Then the United States and several other countries realized the great importance of desalination technology, which will undoubtedly lead to the emergence of a new, very energy–intensive industry.

Prospects for energy supply for an extended period can no longer be associated with hydrocarbon fuel as a primary energy source. Arguments in support of such a statement are well known: fuel is not renewable, its resources are limited, and the cost of production and transportation are increasing. Energy technologies based on it cannot meet current environmental requirements. Moreover, Chernobyl and Fukushima have hardened adverse attitudes toward nuclear energy. Widely believed as the enemy of nature and ecology, is nuclear energy gradually becoming a thing of the past?

All industrial activity of humanity has a technological impact on the environment. If we compare the harmful effects of other technologies on the overall risk to human life and nature generally, then nuclear power plants (NPPs) are the safest of all. Despite the danger of production associated with radiation, during the 50 years in which atomic energy has been developed in the world fewer people have died from radiation than die daily as a result of car accidents. According to scientists the probability of dying as a result of an accident in the nuclear industry is 100 times lower than in a car accident and 1000 times smaller than from heart disease. Many fast-growing countries in Asia and Eastern Europe trying to create conditions for sustainable economic growth along with energy security and reduction of emissions of CO_2 have either resumed the peaceful development of nuclear energy or will do soon.

As the need for energy increases so will the demand for nuclear fuel and its components including raw natural uranium. By 2030 the demand for natural uranium will be 98,000 tons per year if the declared global rate of growth in nuclear capacity is implemented.

Transition to New Nuclear Power Technologies

Modern atomic energy is a mature technology of industrial energy production the reliability, safety, and efficiency of which are confirmed by many years of experience in operating a large number of power plants around the world. Nuclear energy has high potential for technical and economic improvement, but its implementation is hampered as a result of the lack of adequate state support and insufficient investment attractiveness of NPP projects.

To solve the problem of energy shortages and still meet environmental requirements, the part played by nuclear energy needs to increase several times in the twenty-first century. This task can be solved if it is given the status of a national priority program involving government support and ensuring the transfer of knowledge and experience from industry veterans to the new generation of specialists.

The first nuclear reactor symbolically named "Atomic Peace" was created and successfully tested on June 26 in 1954 in the small city of Obninsk on the Protva River not far from Moscow; ever since June 26 has been considered the birthday of the "peaceful atom." Although this is an indisputable historical fact, the United States was the first country in the world to create and test a military molecule on civilians by dropping two atomic bombs of highly destructive power on Japan. On the other hand, Russia was the first country to create a NPP and started the new era of the peaceful atom.

How does a NPP work? At any NPP three reciprocal transformations of energy forms take place: nuclear energy is converted into thermal energy, thermal energy into mechanical energy, and mechanical energy into electric energy.

It looks like this. At the base of the station is a reactor allocated a given volume into which nuclear fuel is loaded and where a controlled chain reaction takes place. Uranium-235 is divided by slow (thermal) neutrons resulting in a considerable amount of heat that is discharged from the core by a coolant, a liquid or gaseous substance passing through its volume. Water is usually used as the coolant. Metal melts in fast neutron reactors (e.g., sodium in a BN-600 reactor). The most complex transformation involves nuclear energy being transformed into thermal energy. The heat collected by coolant in the reactor core is used to produce steam in the steam generator.

History of the Development of the First Nuclear Power Plant

The history of the development of nuclear energy started in Paris in 1896 with the discovery of radioactivity by Henri Becquerel. He discovered that uranium salts emit radiation that (much like X-rays found a year earlier) can pass through cardboard, darken a photographic plate, and ionize the air. The nature and source of this radiation turned out to be a fertile field for research. Then Marie and Pierre Curie discovered polonium and radium (two new elements). The latter was found to be millions of times more radioactive than uranium.

However, a direct way of mastering atomic energy and using it practically was developed by the famous Russian scientist Vladimir Vernadsky, founder of the first nuclear project aimed at helping the state economy. Back in 1908 Vernadsky suggested that since radioactivity can provide heat under natural conditions it can be used for practical purposes. Moreover, new, potent sources of energy should open up as a result of adequately directed efforts. At the time several scientists were already engaged in radioactivity in Russia and Europe. However, it was Vladimir Vernadsky who showed radium's importance in practical use. In the autumn of 1910 a Radium Commission was set up at the Academy of Sciences at the suggestion of Vernadsky, which also included top executives such as Prince Golitsyn and the very prominent academicians Beketov, Karpinsky, and Chernyshev. On December 29, 1910, at the ceremonial annual meeting of the Russian Academy of Sciences, Vernadsky read out a program report entitled *Tasks of the Day in the Field of Radium*, in which he presented a whole program of geological and technological research and detailed a methodology aimed at finding uranium ores and mastering the energy of atomic decay. Vernadsky said in his report: "… Now before us, in the phenomena of radioactivity, the sources of atomic energy are discovered that are millions of times greater than all those sources of power that were drawn to the human imagination" (Vernadsky 1954). Meanwhile, in England the British physicist Ernest Rutherford discovered the alpha and beta radiation of uranium nuclei, and identified the proton as an independent elementary particle. He was the first to carry out an artificial nuclear reaction that consisted of converting nitrogen into radioactive fluorine, which in turn gave rise to stable oxygen. The main scientific achievement of Rutherford was creation of a revolutionary model of the atom. According to this model the mass of a particle is not smeared over space and almost all of it is concentrated in a small area that carries a positive charge. He drew this conclusion based on the results of experiments on the passage of alpha particles through gold foil. Based on the Rutherford model the famous scientist Niels Bohr in 1913 proposed a quantum model of a hydrogen-like atom known as the Bohr atom. Rutherford created an extremely productive scientific school. Experimental studies by his students and staff led to the discovery of the neutron by James Chadwick, the identification of helium-3 and tritium by Marcus Olyphant and Paul Hartek, the production of superstrong magnetic fields and development of the superfluidity of liquid helium by Russian scientist Kapitsa. Among Rutherford's students was the founder of quantum mechanics Niels Bohr and one of the leading developers of Soviet nuclear weapons Khariton. During World War I, Rutherford expressed the hope that atomic energy could not be mastered until people learned to live in peace with each other. As a motto for his baronial coat of arms Rutherford chose a quote from Lucretius *Primordia Quaerere Rerum* ("to seek the first principles of things"). He followed this principle in his scientific work all his life. A little over a year later he died, the very year that Otto Hahn and Fritz Strassmann discovered the neutron-induced fission of uranium, which opened the way to nuclear weapons and military nuclear power.

The path from the first nuclear boiler to the NPP was almost four times longer than the road to the atomic bomb. The first industrial plants that were set up involved a controlled chain reaction in which steel reactors were divided for the production of plutonium-239 (Pu-239), which is formed during the absorption of neutrons by uranium-238. After that there were small power plants for submarines. In 1951, at the American experimental station in Idaho, scientists even managed to get some electricity—enough to light up four bulbs.

The problem was that completely different temperature and power conditions were required to operate a full-fledged power plant comparable in power to thermal power plants (TPPs). To turn four light bulbs into millions it was necessary to have not just a different quantity but a different quality of nuclear power. It was essential to organize heat exchange at high heat fluxes and high pressures—questions that have been poorly studied. Fuel cells in the core had to operate at high temperatures without destruction and structural materials had to withstand enormous radiation loads.

In 1950 the Soviet Union decided to build an experimental reactor. Several design options were proposed, but the final draft approved by scientists Kurchatov and Dollezhal was supported. This involved building a uranium–graphite reactor with tubular fuel elements where non-boiling water was used as the coolant. The design capacity of the reactor was based not on theoretical considerations but on particular circumstances. At that time, when there were general post-war shortages even in priority areas, it was often necessary to be content with available resources. However, when searching for a turbine unit in Moscow they stumbled upon an old, out-of-service turbine of small capacity. Although only about 6 MW, it was quite suitable for an experimental NPP. The characteristics of this turbine ultimately determined the power of the world's first NPP.

The whole process of building a NPP, which started in September 1951, was a series of experiments and tests. Analyzing potential emergencies by looking at specific parameters of the reactor it was realized that the most dangerous thing would be to fill the masonry with water. If the cooling channel was broken the multiplication factor of neutrons would increase and the power begin to increase. Several laboratories in post-war Russia got engaged in the development of fuel rods (fuel elements). They contained nuclear fuel and were located in the most dangerous zone of the reactor. First, the fuel rods were made in the form of steel tubes on which uranium bushes were mounted. When pure uranium was replaced by alloy with 9% molybdenum things went better with service life increasing to several hundred hours. However, the most successful solution was to disperse a uranium–molybdenum alloy in a magnesium matrix. When tested this design dealt with heat fluxes. In early May 1954 the scientists started loading the core with fuel. The first complete batch of nuclear fuel contained 546 kg of uranium with 5% enrichment of uranium-235. The tests continued for a month and a half, and then on June 26, 1954 at 5.45 p.m. steam was supplied to the turbine and the first NPP in the world received an industrial load with an electric generator power of 1.5 MW. The launch of the first NPP gave rise to hopes for clean energy with virtually unlimited resource potential.

A couple of years after the Soviet Union set up the first NPP, in 1956 in Great Britain the Calder Hall NPP with a capacity of 45 MW and in 1957 in the United States the Shipping Pork NPP with a capacity of 67 MW started their operations. Over the years nuclear power has taken on critical importance in the global energy balance.

Indeed, innovative approaches to the design, construction, and operation of NPPs is a requirement of our time. The development of fast-neutron reactor technology allows us to solve several essential tasks such as ensuring the safety of NPPs and the efficient use of nuclear fuel.

Moreover, "fast energy" is key to solving many problems in the nuclear industry related to national security and environmental protection. First, as a result of greater utilization of uranium in breeders its extraction and storage have some environmental impacts.

Measures have been put in place internationally to stop the mining of uranium for the foreseeable future. This is because it can be produced from spent nuclear fuel (SNF) and depleted uranium hexafluoride (DUHF), the reserves of which are more than sufficient. Second, the environmental impact of radioactive waste management is now reduced. Radioactive waste requires the arrangement of expensive storage facilities to ensure its isolation from the environment for a sufficiently long time for it to decay. Breeders can dramatically reduce the time needed to isolate it and reduce the potential danger. Third, plutonium obtained from the processing of SNF from thermal neutron reactors can also be disposed of in breeders.

The opposite approach (i.e., storage and final disposal of plutonium as radioactive waste) requires special measures to ensure safety and consequently incurs high costs. The advantages of fast reactors are apparent. Today a new market is opening up in which countries with nuclear technology can supply technology and high-tech equipment. At the same time there is a risk of losing the "fast" technology race to developing states such as China or India. Therefore, it is necessary to accelerate the development of this area in the nuclear power industry.

Atomic energy in its present form was created during the Cold War and is its legacy. The main task of nuclear energy at that time along with the production of energy was the creation and accumulation of nuclear materials. This was the reason existing NPPs based on the uranium–plutonium cycle were created. Everything else—such as the treatment of radioactive waste and SNF, their storage, the decommissioning of NPPs themselves at the end of their service lives, the risk of proliferation of nuclear materials, and other environmental issues—was of secondary importance. It was also believed that NPPs would be very reliable and safe stations.

Over the past 20 years the number of orders for new nuclear installations worldwide has declined. The reasons include public resistance to the construction of new NPPs and the presence of excess capacity of power units in many energy markets, mainly due to the development of competition in the electricity market and low economic performance of many existing nuclear facilities. At the same time many developing countries in Southeast Asia, the Middle East, and Africa are showing a keen interest in developing nuclear energy in their countries. The decision on whether to start or continue to create atomic energy will ultimately depend on the choice of tasks facing the state in question. No other type of energy or industry has grown at such a fast pace as nuclear energy. In 1954 the world's first 5-MW NPP was commissioned, and as of January 1, 1980 about 80 NPPs with a capacity of 15,000 MW had been approved in the world.

The global energy crisis of the early 1970s caused by the growth in electricity demand and the oil embargo imposed by the Organization of Petroleum Exporting Countries (OPEC) countries in 1973–1974 led to an increase in energy prices and gave additional impetus to the development of nuclear power throughout the world. It also contributed to public opinion being firmly convinced that nuclear power would be the only real alternative source of energy for the foreseeable future. Therefore, until the mid-1980s atomic energy was considered the most promising source of energy, a means of overcoming the energy crisis. In only 20 years (from the mid-1960s to the mid-1980s) the global share of electricity produced at NPPs increased from almost 0 to 17%, and in some countries it became prevalent (about 80% in the case of France).

Advantages of Nuclear Power Plants (NPPs)

The main advantage NPPs have over coal and gas energy plants is they do not need to be near energy sources as a result of the small amount of nuclear fuel used. This allows building NPPs without being tied to resource generation since the cost of transportation is not high because of the low volumes. Suffice it to say that 0.5 kg of nuclear fuel produces the same amount of energy as the burning of 1000 tons of coal.

The second significant advantage of NPPs is their relative ecological purity. For the sake of comparison, at TPPs total annual emissions of harmful toxic and greenhouse substances (such as sulfur dioxide, nitrogen oxides, carbon oxides, hydrocarbons, aldehydes, and ash dust) per 1000 MW of installed power are approximately 13,000 tons per year at gas-powered and about 165,000 tons at coal-powered stations. At NPPs such emissions are absent. For the oxidation of fuel a TPP with a capacity of 1000 MW consumes 8 million tons of oxygen per year, whereas NPPs do not consume oxygen at all. Moreover, coal always contains natural radioactive substances that are almost entirely released into the environment when fuel is burned. At the same time the specific radioactivity of TPP emissions is several times higher than that of NPPs.

The situation regarding further development of NPPs dramatically changed after the Chernobyl disaster in April 1986 and after the accident at the Japanese Fukushima 1 NPP in March 2011, in particular. These events signaled the demise of the myth surrounding safe nuclear power engineering and bolstered the belief that reactors are always dangerous. The main environmental hazard of NPPs was attributed to the possibility of accidents. Although the probability of such accidents at modern NPPs is low, it is impossible to eliminate them.

By transitioning to a different fuel cycle the construction of fast-neutron reactors using plutonium-239 (Pu-239) or thorium-232 (Th-232) instead of uranium as fuel has the potential to dramatically affect the nuclear power industry. Such innovations are aimed ultimately at reducing the amount of uranium mining, which could adversely affect prices. However, it should be taken into account that mass application of these technologies may well be possible soon and that complete restructuring of the modern nuclear power system will entail enormous material costs.

In the twenty-first century five basic requirements have been imposed on nuclear power: safety, economy (competitiveness compared with other energy technologies), non-proliferation, management of spent fuel, and management of radioactive waste. The search for safe and economical ways to develop nuclear energy can be divided into two main areas. The first implies that all efforts should be focused on eliminating public distrust of the safety of nuclear technologies. For this to happen it is necessary to develop new reactors that are safer than current light-water reactors. The two types of reactors that have been presented are the technologically extremely safe reactor and the modular high-temperature gas-cooled reactor. The prototype modular gas-cooled reactor was developed in Germany, the United States, and Japan. Unlike a light-water reactor the design of a modular gas-cooled reactor is such that its safety is ensured passively without direct actions by operators or an electrical or mechanical protection system. In technologically extremely safe reactors a passive protection system is also used. The future of these options is uncertain because of their unpredictable cost, development difficulties, and the uncertain future of the nuclear power industry itself. The second direction assumes that developed countries will need new power plants. The primary task is to stimulate investment in nuclear power.

In addition to these two prospects for the development of atomic energy a completely different point of view has arisen that places hopes on complete utilization of the energy supply, renewable energy resources, and energy saving. According to supporters of this point of view the electricity saved will be enough to do without all the existing NPPs if advanced countries switch to the development of more economical household appliances, heating equipment, and air conditioners. The observed significant decrease in electricity consumption shows that the economy can be an essential factor in limiting the demand for electricity. The crisis of the environment urgently requires transition to a new energy technology that is not related to the combustion of hydrocarbon fuels. At present only nuclear power engineering based on fission reactors can be considered an alternative energy source.

To develop coordination in the peaceful use of atomic energy the International Atomic Energy Agency (IAEA) was founded in 1957. The IAEA is an influential global organization with its headquarters in Vienna (International Vienna Center). The agency was created as an independent intergovernmental organization within the United Nations. With the advent of the Treaty on the Non-Proliferation of Nuclear Weapons (NPT) its work took on special significance since the NPT made it mandatory for each State (party to the Treaty) to conclude a safeguards agreement with the IAEA.

The goal of the agency's work in any State is to make it clear that work in the peaceful nuclear field should not switch to military objectives. By signing such an agreement a State guarantees that it does not conduct studies of service orientation, hence this document is called a compromise on guarantees. At the same time the IAEA is a purely professional body. It cannot give a political assessment of the activities of a State. The IAEA does not have the right to speculate since it only works with facts currently available and bases its findings solely on the tangible result of inspections. The IAEA safeguards system cannot physically prevent the diversion of nuclear material from peaceful to military purposes, but only makes it possible to detect the diversion of safeguarded material or the misuse of a secured facility and initiate consideration of such facts at the United Nations. At the same time the conclusions of the agency are incredibly cautious and correct.

The functions of the agency include:

1. Formation and application of a system of guarantees that civilian nuclear programs and developments will not be used for military purposes.
2. Development, establishment, and adaptation of health and safety standards.
3. Encouraging research and development on the peaceful uses of atomic energy.
4. Encouraging the exchange of scientific advances and methods.

Hence the advantages of nuclear power can briefly be formulated as:

1. Colossal energy consumption of the fuel used: enriching 1 kg of uranium to 4% at full burnout emits energy equivalent to burning about 100 tons of high-quality hard coal or 60 tons of oil.
2. Possibility of fuel reuse (after regeneration): fissile material (uranium-235) can be used again (as opposed to ashes and slag from fossil fuels); with the development of fast-neutron reactor technology a transition to a closed fuel cycle is possible in the future, which means the complete absence of waste.
3. Nuclear power does not contribute to the greenhouse effect: every year NPPs in Europe make it possible to avoid emissions of 700 million tons of CO_2; in addition, operational Russian NPPs annually prevent discharge into the atmosphere of 210 million tons of CO_2; thus, the intensive development of nuclear energy can be indirectly considered one of the methods to combat global warming.

However, only about 18% of today's world electricity production comes from NPPs. The industry ranks third after coal and hydropower. In general, the world nuclear power industry today includes 448 nuclear reactors in 31 countries and produces about 370 GW of electricity in total. Most of the power plants are operated in the United States (99), France (59), Japan (55), Russia (31), and China (38). The total generating capacity of NPPs is over 392 GW.

General Characteristics of the Nuclear Industry Worldwide

Although there are 388 nuclear power reactors in 31 countries with a total capacity of 333 GW currently operating in the world, the industry is currently experiencing a recession. The share of nuclear power in global electricity production decreased from 17.6% in 1996 to 10.8% in 2013.

The Russian state corporation ROSATOM is a global leader in the peaceful atom. Russian company TVEL supplies fuel to 73 reactors (17% of the world market). France is a world leader when it comes to the part NPPs play in the national electricity production. Nuclear power in France is developing in a planned and sustainable manner. France operates 59 nuclear power units with a total electrical capacity of about 70 GW, which represents 78% of the country's total electricity. At the end of 1997 a reactor—belonging to the third generation of high-security reactor installations—using pressurized water (PWR) with an electric capacity of 1450 MW was commissioned in France. One of the creators of the first Soviet centrifuges Konrad Zippa subsequently worked with the Anglo-Dutch concern Urenco and developed a new model of the centrifuge for them. This was smuggled to Pakistan by Abdul Qadeer Khan, a Pakistani who later gave it to Iran.

Two-thirds of the reactors currently under construction are in China, India, and Russia. The average age of operating reactors is 28.5 years. The oldest is located in Switzerland and has been working for 45 years. A total of 153 reactors have been closed. The average age of closed reactors is 23 years. Most currently operational nuclear power units use light-water reactors representing a little over 80% of the total number of blocks. Heavy-water installations represent 10%, gas-cooled 4%, and reactors with water cooling and a graphite moderator 4%. All these reactors operate on thermal neutrons. There is also one fast-energy power reactor using the sodium coolant BN-600 in the world (in Russia).

Most NPPs are in the United States where they provide about 20% of the country's electricity. In second place is France with 58 units and Japan was in third place before the accident at the Fukushima NPP. There are 10 NPPs in operation in Russia (33 power units). The world leader when it comes to using NPPs is France. Its 58 units produce about 80% of the country's electricity and have a total capacity of about 64 GW. The most powerful NPP in the world is Kashiwazaki Kariva (Japan) with a capacity of 8200 MW (7 BWR-type reactors with an installed capacity of 110–1356 MW). In Europe the most powerful is Zaporizhzhya NPP (Ukraine) with a capacity of 6000 MW (6 VVER-1000 reactors).

Nuclear Resource Base

The main resource base of nuclear power engineering today is natural uranium. Uranium-235 comprises 0.7% natural uranium and is used as fuel at NPPs. Rich uranium ores are distributed very unevenly in the world. Over 20% of the world's known resources are concentrated in Kazakhstan. For use at NPPs mined uranium is enriched in uranium-235 by between 3 and 5%. When enriched by more than 20% it can be used as a material for creating nuclear weapons. Although this is potentially applicable to nuclear energy thorium, so far it has not found any practical application.

The World Nuclear Association estimated the reserves of recoverable uranium as approximately 5,404,000 tons. Maintaining the current number of nuclear power units and the installed capacity of NPPs these reserves will be enough for 90–100 years. The lifetime of a modern NPP is 60 years. When it comes to uranium reserves Australia is in first place with 1,673,000 tons. In second place is Kazakhstan with 651,000 tons. In third place is Canada with 485,000 tons. Russia, excluding foreign deposits wholly or partly owned by ROSATOM, is in fourth place with 480,000 tons or about 9% of global uranium reserves.

In terms of uranium mining Kazakhstan continues to occupy first place for the third year in a row with 19,451 tons in 2018. The former leaders Canada and Australia are in second and third places, but their production levels are steadily decreasing. In 2018 Canada produced 9145 tons of uranium, Australia 5983 tons, and Russia 2993 tons. The uranium market is monopolized to a high degree. The top five countries (Canada, Australia, Kazakhstan, Niger, and Russia) account for almost three-quarters of world uranium mining. The outlined shortage of nuclear fuel and monopolization of its market will certainly lead to higher prices for enriched uranium and accordingly to the economic performance of NPPs deteriorating. Since uranium reserves are limited further development of nuclear energy with thermal neutrons is also limited.

Nuclear Proliferation Risk

Nuclear power in its present form was created during the Cold War to produce nuclear weapons, and is its legacy. The atomic ministries of various countries received energy as a by-product of nuclear fission, but at that time did not consider environmental pollution. The vast majority of today's nuclear power makes use of the uranium–plutonium fuel cycle, which brings about the production of plutonium-239 used to obtain nuclear weapons. All reactors including thermal ones produce plutonium-239 in the course of their work. By processing recycled nuclear fuel to produce useful nuclear materials this will make plutonium-239 more affordable for those who would like to possess nuclear weapons. According to some estimates 1633 tons of plutonium was produced worldwide in 2000 consisting of 1384 tons of civilian plutonium and 249 tons of military plutonium. In addition to military plutonium a large amount (up to 67.5%) is contained in NPP SNF and can be separated if necessary.

In 2000 the United States accounted for 85 tons and Russia about 100–165 tons (exact data have never been published). The risk of proliferation of nuclear weapons increases manifold when states decide to build hundreds of new uranium–plutonium NPPs in Third World countries. At present, with the rise of terrorism the risk for proliferation of clean or dirty nuclear weapons has increased. The number of people willing to sacrifice their lives for religious beliefs over the past decade has grown more than 10 times illustrating just how serious the situation is in the world. Therefore, one of the most critical problems facing the modern world is the development of new technologies in the nuclear power industry that can make it possible to reduce the risk of nuclear proliferation significantly.

Reference

Vernadsky, V. (1954). *Collection of the Selected Research.* Moscow: Publishing House of the Academy of Sciences of the USSR.

3

Geopolitical Aspects of Nuclear Power

"Nuclear World" as a New Phenomenon of the Globalization

Much time has passed since the advent of nuclear weapons and the first nuclear power plants. During this period, the number of nuclear countries increased significantly, and the number of states with enterprises in the nuclear industry and nuclear energy increased. At first, work on the development of nuclear energy for peaceful and military purposes was carried out under national flags. The countries independently built up and increased their military nuclear potential, built nuclear power plants, equipped ships, and submarines with nuclear power plants. All this was done in the interests of ensuring the military security of the state, as well as the economic development of the country.

Actions aimed at changing the state of nuclear forces of one state forced other countries to counteract by changing the capabilities of their nuclear forces actively. This circumstance was one of the reasons for the appearance of the nuclear arms race. That led to the understanding that the world with nuclear weapons is permeated by bilateral and multilateral nuclear ties, which firmly united countries among themselves, making them nuclear-dependent.

D. Basu and V. W. Miroshnik, *The Political Economy of Nuclear Energy*,
https://doi.org/10.1007/978-3-030-27029-2_3

The technological dependence of states in the field of atomic energy is continually growing. Some countries become suppliers of equipment and fuel for nuclear power plants, others—only their consumers, and still others—processors of radioactive waste from NPPs. The consolidated activities of states and the international differentiation of labor in the atomic energy sector reinforce the technological nuclear dependence of countries on each other.

Interdependence is also growing between the military and the peaceful sectors of the nuclear sphere. Thus, within the framework of a single technological cycle (at the same power plants), the production of filling ammunition and fuel for nuclear power plants is carried out; at the same radiochemical plants, nuclear ordnance, and spent nuclear fuel from nuclear power plants are processed.

There is an increase in the interdependence of nuclear countries within the framework of military-political alliances. This process is primarily associated with the creation of coalition nuclear forces. So far there is one military-political bloc, the North Atlantic Treaty Organization (NATO), in the world, in which coalition strategic offensive forces are consisting of the strategic attacking troops of the USA, Great Britain, and France. For the everyday use of the NATO coalition forces, a collective nuclear strategy has been formed, the basis of which is the overall atomic planning. As a result, allied nuclear countries have a very high level of nuclear dependence. These activities and interdependencies are forming the globalization process in the nuclear sphere, which is manifested in the structuring of the nuclear field and the formation of specific nuclear bonds between countries.

Under the influence of the globalization process in the nuclear sphere, individual states with nuclear weapons and nuclear technologies formed a qualitatively new international system in the twenty-first century—the geopolitical order of nuclear-dependent states with inherent systemic properties such as stability all countries with atomic challenges, dangers, and threats.

Changes in the quantity and quality of subjects of the nuclear world, nuclear connections and relations occurring under the influence of internal and external factors, lead to the transformation of the nuclear world, changing its system properties. At the same time, the nuclear world as a geopolitical system moves from one qualitative state to another, creating new nuclear dangers and threats.

By the nature of the structure and behavior of the system forming the nuclear world can be attributed to the socio-geopolitical. Even a brief listing of the geopolitical aspects of the system allows us to classify it as geopolitical. The following circumstances indicate its geopolitical side:

- The scale of the system, covering all nuclear states, countries with nuclear and energy enterprises, international governmental and non-governmental organizations dealing with nuclear issues, and other geopolitical actors;
- Each nuclear state and country possessing nuclear industry enterprises form national nuclear geopolitics. The combination of national nuclear geopolitics of individual states creates a specific geopolitical environment, which largely determines the lines of the geopolitical behavior of countries;
- The fact of the presence of nuclear weapons in the state or the possibility of their soon occurrence changes the political attitude toward it on the part of other countries. Today, nuclear weapons are viewed by politicians primarily as a "political weapon" that can significantly influence the geopolitics of the surrounding states;
- Multilateral and bilateral international treaties and agreements in the nuclear sphere are of a pronounced geopolitical nature.

As for the social aspect, it is quite evident that nuclear geopolitics conducted by states affect all spheres of social life both within the state and the world community such as interstate relations, social security of the population of nuclear countries, corporate social relationships, and public organizations. Also, the Treaty on the Non-Proliferation of Nuclear Weapons (NPT) has begun to determine the international social status of a nuclear state. The presence of nuclear weapons in some countries and the lack of them in other socially divide nations into two camps—nuclear and non-nuclear countries (in terms of nuclear weapons). And, as a consequence of this, a global social polarization of relations between nuclear and non-nuclear countries arises, which is gradually increasing. All this allows us to consider the phenomenon not only as a geopolitical but also a social system.

Thus, it can be stated that a new geopolitical system has emerged, which it is advisable to give the name "nuclear world" (NM). It should be noted that the new phenomenon such as "nuclear world" does not fit into the framework of traditional concepts: "nuclear club," "nuclear powers," "nuclear states," and "nuclear countries." A qualitatively new global geopolitical substance emerged, which is a geopolitical system that simultaneously encompasses the global, regional, and national nuclear levels, as well as the most significant segments of the nuclear sphere: nuclear energy and nuclear weapons complex.

Structure of the Nuclear World

The first component of the nuclear world. That includes countries that are legally producing nuclear weapons from the Treaty on the NPT such as Russia, the USA, Great Britain, France, and China, which are traditionally united by the concept of a "nuclear club." The countries of the nuclear club in their foreign policy take into account the presence of nuclear weapons from each other, which allowed them to develop several consolidated measures to identify nuclear threats, neutralize or localize them. A feature of the nuclear club is to divide the members of the club according to interests into several conditional groups.

Therefore, the USA, Great Britain, and France unite in a separate group. They are members of the NATO bloc, form a coalition nuclear strategy within the bloc, and advocate their group interests.

Russia and China, as relatively independent actors, occupy a separate position in the nuclear club, defending their interests, which in many respects do not coincide with the benefits of NATO countries.

The second component of the nuclear world. It consists of a group of countries that are illegally producing nuclear weapons, the countries whose status is derived from the provisions of the NPT. A characteristic feature of the states of this component of the nuclear world is uncertainty in the political, military, and economic contradictions relations within the element that gave rise to an atomic confrontation between states. In this component, there are risks of using nuclear weapons in armed conflict, as well as manifestations of nuclear terrorism and nuclear blackmail.

The third component of the nuclear world. These are shadow nuclear states. They do not officially recognize themselves as nuclear powers. Nevertheless, these states are possessing atomic weapons, and they consciously hide this fact from the world community. Analysis of numerous sources gives grounds to refer to this component of the nuclear world such state as Israel. In addition to Israel, there may be other shadow nuclear states. With the continued proliferation of nuclear weapons and the growth of international instability, there is a risk of increasing the number of the country in this component of the nuclear world. For example, we cannot exclude the replenishment of this component by several Islamic states.

Shadow nuclear states have their distinctive features. In particular, the views of their political leadership on the preparation and use of nuclear forces are not transparent for other members of the world community. They have a significantly increased risk of erroneous assessments about their neighbors' military preparations and making wrong decisions to counter these preparations. All this indicates that the presence of shadow nuclear countries makes the nuclear world less predictable.

The fourth component of the nuclear world. That includes the so-called "border" nuclear countries such as North Korea, which have the technology to create nuclear weapons, and have means of delivery, but for political reasons do not develop nuclear weapons and do not form national nuclear forces. To date, North Korea has conducted several explosions of nuclear devices and is armed with aircraft, ballistic, and cruise missiles, which by their characteristics are capable of carrying nuclear weapons. According to these signs, North Korea can no longer be attributed to non-nuclear countries. However, it cannot be considered a nuclear state for the time being, since the country does not officially have nuclear weapons (a nuclear weapon that is technology combined with a carrier) and no nuclear forces were created yet.

A characteristic feature of the border countries is the geopolitics of balancing on the edge of status between a nuclear and non-nuclear power. In such position, a state is capable of taking political, economic, and military benefits from its border position for a long time. Perhaps for this reason, in the twenty-first century, other states can replenish the component of the nuclear world as nuclear border countries.

The fifth component of the nuclear world. That includes the so-called "threshold" countries (which are on the verge of creating nuclear weapons). Today they can be divided into two subcomponents: threshold first countries and subsequent queues. A distinctive feature of the first countries of the limit is the desire of their political leadership to possess nuclear weapons. Sometimes this desire is realized through clandestine operations for the illegal extraction of nuclear materials and technologies.

A blurred line between the shadow, border, and threshold first countries is characteristic of the modern nuclear world. Likely, some threshold first countries can already be attributed to the shadow or nuclear border states.

The next countries to the threshold countries are states that, according to the level of industrial development, can create their nuclear charges, such as Argentina, Brazil, South Korea, Japan, and more than ten other countries.

The danger of the existence of threshold countries lies in the fact that they create a powerful potential for increasing the number of illegal shadows and border nuclear states, introducing uncertainty in the future development of the nuclear world.

The sixth component of the nuclear world. It includes states with existing nuclear power industries. There is a danger that such enterprises will be chosen as a target for attack by the aggressor state, and the result of the attack will be equivalent to nuclear effects. Also, the nuclear industry can be subject to a terrorist attack. Such states pose a significant danger from nuclear accidents and environmental disasters of global and regional scale. Taking into account that there are more than 400 nuclear industry enterprises in the world, it can be concluded that it is necessary to unite states with nuclear industry enterprises into a particular nuclear risk group.

Seventh component of the nuclear world. That includes non-governmental organizations, including terrorist organizations, religious sects, and extremist groups, possessing some nuclear materials and radioactive substances that can be used to conduct terrorist acts.

Eighth component of the nuclear world. That includes international governmental and non-governmental organizations, national institutions, and movements that regulate or influence nuclear relations. For example, these include:

1. United Nations Organization (UN) and its structural divisions—the International Atomic Energy Agency IAEA and the United Nations Educational, Scientific and Cultural Organization UNESCO;
2. International governmental and non-governmental organizations in the field of the use of nuclear energy such as
   - International Commission on Radiation Units and Measurements (ICREI);
   - World Association of Nuclear Operators (WANO);
   - International organizations and research centers (European Nuclear Society, European Atomic Forum, International Energy Agency (IEA), Organization of Nuclear Energy Producers, US Atomic Energy Commission, European Laboratory of Particle Physics, Office of Nuclear Research Safety of the Commission of European Communities, etc.);
   - National organizations and research centers;
3. International and domestic organizations investigating the effects of radiation on human health and the environment: World Health Organization, International Radiation Protection Association, International Association of Radiation Research, International Commission on Radiation Protection, United Nations Scientific Committee on the Effects of Atomic Radiation, United Nations Environment Program (UNEP), International Agency for Research on Cancer, International Union of Radio Ecologists, European Society of Radiobiologists, National Research Centers.

Ninth component of the nuclear world. That includes the International Nuclear Information System, established in 1970 by the International Atomic Energy Agency (IAEA) in collaboration with the member states of the agency. The organization is designed to collect, process, and disseminate information on the problems of nuclear energy, nuclear technology, and radiology. One of the sources of information replenishment in the system is specialized publications produced in various countries of the world.

The nuclear world has many links, forming multiple connections and relationships; these existing nuclear links should be divided into several conditional levels.

The first level of the nuclear world is bilateral (paired) nuclear communications. They arise between nuclear countries when nuclear weapons of one state have technical capabilities—the range and accuracy of delivery of nuclear warheads—have real capacities to hit strategic targets on the territory of another, non-union nuclear state, and the same capabilities as nuclear weapons of the opposite state. In this case, the capabilities of the nuclear weapons of the two opposing states are the basis of the nuclear connection, which forms a polarized pair of two nuclear countries with normal bilateral nuclear relations. The degree of polarization of ties in a couple is not constant; it changes with the situation. The main link in the pair relationship is the qualitative and quantitative ratio of nuclear forces—it becomes the subject of close attention from the political and military leadership of the paired states. The change in the balance of forces in favor of one state almost always causes concern on the part of the pair state and forces the latter to take retaliatory measures.

It should be noted that in the nuclear world, many states simultaneously form paired ties with several countries at once. For example, Russia creates pairs with the USA, France, Britain, China, India, and Pakistan; India—simultaneously with Pakistan, China, and Russia. Thus, almost all nuclear countries are covered by paired nuclear links. It can be assumed that paired nuclear connections form the initial "framework" of the modern nuclear world.

The second level of the nuclear world is the nuclear geopolitics of the states. It is formed on the basis of awareness of the essence of the vital interests of the state in the nuclear field; the political leadership's understanding of the diversity of bilateral and multilateral nuclear ties among countries; forecasting the nature of changes in nuclear threats for the near and distant future of the development of interstate relations. The nuclear geopolitics of various states are closely intertwined, thus, creating a conflict for the subjects of the nuclear world.

The third level of the nuclear world is the coalition of the global nuclear geopolitics of states. This level is formed on the basis of allied obligations between countries as a result of their blocks' policy. For example, within the framework of NATO, countries such as the USA, Great Britain, and France are developing a global coalition in nuclear geopolitics, which has a significant impact on the alignment of forces in the modern nuclear world. In the future, we can assume the emergence of new coalitions in global nuclear geopolitics, which can significantly complicate relations between states.

The fourth level of the nuclear world is the coalition of the regional nuclear geopolitics of the states, which is formed within one region. Such geopolitics is aimed at the timely detection of nuclear threats at the local scale, the neutralization or localization of such risks. However, today, such regional nuclear geopolitics is practically not formed in any region of the world. Nevertheless, this does not mean that it cannot be built shortly when, under the influence of increasing nuclear hazards and threats, regional communities realize the need to consolidate efforts and begin to form local institutions designed to develop and implement regional nuclear geopolitics.

The fifth level of the nuclear world is nuclear geopolitics of international institutions. It is intended to regulate nuclear interstate relations in the interests of neutralizing or localizing threats of a global scale. Today, within the framework of the UN and its structural subdivisions—the IAEA and UNESCO—the foundations of such international nuclear geopolitics are already being formed. Its content is the ongoing work of organizations to prevent the proliferation of nuclear weapons, arms control, and reduce the level of nuclear danger.

The sixth level of the nuclear world is a regional nuclear geopolitics. It is based on multilateral and bilateral treaties in the nuclear field, and the content is a set of forbidden, restrictive, permissive, and other measures aimed at banning nuclear weapons tests, establishing a non-proliferation regime, reducing the threat of nuclear war, preventing unauthorized use of nuclear weapons, nuclear terrorism and nuclear blackmail, as well as measures to prevent accidents and disasters at the nuclear energy enterprises.

The seventh level of the nuclear world is the cooperation of states in the field of nuclear technologies. Today, the economic development of any country is unthinkable without nuclear technology. The energy crisis is forcing countries to look for ways out of it. Therefore, cooperation in the field of nuclear energy is continuously expanding, making some states dependent on others on the supply of fuel and equipment for nuclear power plants, training of nuclear specialists, processing of spent nuclear fuel, disposal of radioactive waste, etc.

Thus, the analysis and enumeration of the primary nuclear connections make it possible to judge that a large variety of nuclear relations is forming the nuclear world, generating a complex communicative environment.

The structural-territorial construction of a nuclear world is an essential aspect of the knowledge of the geopolitical system. Attention should be paid to the fact that the subjects of the nuclear world are unevenly located in different regions.

System Properties of the Nuclear World

In system analysis, special attention should be paid to the identification of systemic (emergent) features of the object of knowledge. The main system properties of the nuclear world are as follows: (1) the stability of international relations in nuclear materials; (2) strategic stability of the nuclear materials; (3) nuclear safety as a property that characterizes the state of protection of individuals, societies, nations, and the world community from possible nuclear hazards and threats associated with the production, testing, operation, transportation, storage, and disposal of nuclear weapons, nuclear materials, and radioactive substances.

Global and Regional Levels of the Nuclear World

The nuclear world at the global scale is composed by all nuclear countries such as legal, illegal, shadow and border threshold states, and states possessing nuclear power and developing nuclear industry, international organizations dealing with nuclear issues, as well as non-state formations that carry out plans for using nuclear materials in various purposes.

The nuclear world at the regional level represents the subjects of the nuclear world within one region. Today, by the number of nuclear countries, the most saturated part is Central Asia (CA), where there are four nuclear states (Russia, India, Pakistan, and China), threshold states (Iran), as well as multiple terrorist groups seeking to possess nuclear weapons.

4

Economic Aspects of Nuclear Power

Nuclear Power and Technological Sovereignty

Technological sovereignty is the ability of the state to provide scientific, technical, and industrial development to create and maintain on its territory its technologies and infrastructure sufficient to guarantee the independence of its policies, economy, and defense capability from foreign technologies in critical, vital areas.

Political scientists say a lot that in recent years the arrangement of global forces has changed: for example, the United States and Europe are seriously losing ground, and China and Japan are rapidly catching up. It is not only about the military or economic aspects, but rather about a technological breakthrough. To strengthen its position, Russia also cannot do without breakthrough technologies. The key to this door could be the nuclear industry.

The concept of sovereignty—political, economic, financial, technological—is closely related to the ideas of world hegemony and the poles of power. Any hegemonic power seeks to gain control in different areas through a built-in system of relations and mechanisms for the distribution of resources and production factors.

D. Basu and V. W. Miroshnik, *The Political Economy of Nuclear Energy*,
https://doi.org/10.1007/978-3-030-27029-2_4

For example, the neoliberal doctrine adopted by American politicians implies maximum control over advanced technologies from the United States and their corporations, as well as setting their standards as international while simultaneously globalizing trade and free access to consumer markets around the world.

It is worth noting that, despite the success of Globalization in the past two decades, neoliberalism has slowed down the development of several countries and regions, increased inequality and imbalances, and caused recurring economic and political crises.

As long as the world lives in the existing financial and economic system, the dominant power determines the rules of the game and monitors their implementation by other countries. But the world is changing. It is quite natural that the hegemonic approach to geopolitics and global governance is met with resistance from powers seeking to preserve sovereignty.

As Wallerstein (1974) noted in his theory of world-system analysis, "*the state loses hegemony not because it is weakening, but because others are gaining strength*" Globalization in the American way has reached its natural limit, with China and Asia as a whole becoming the primary beneficiaries of world development. The Asian region demonstrates high economic growth and active mobilization opportunities. Ahead is the struggle for strategic dominance in politics, economics, energy, and technology.

The main direction in the field of high technologies, on which it is necessary to focus, is just nuclear power engineering. It is the basis of national security and nuclear parity; it is the energy of the present and the future, which will help humanity to get rid of greenhouse gases and climate disasters, providing it with clean sources of energy.

According to the research of the World Nuclear Association (WNA), it is Russia that is now the world leader in nuclear power engineering: in the field of technologies for atomic energy, in the development of new reactor models, including fast neutrons. The efficiency of power generation in our country is also very high.

Economic Perspectives of Global Nuclear Energy

Aristotle came up with the concept of "energy" to denote active human activity that transforms the world around. Nowadays, the degree of progress and the level of development of a civilization are usually measured by the amount of energy consumed by humanity, as well as the amount of information available to it. The need for power, particularly nuclear energy, in the last 100 years, is growing in proportion to the square of the population of the Earth. Experts predict a 60% increase in energy consumption over the next 15 years and a 50% increase over the 25-year interval. The share of developed countries by 2040 will account for 35% of the total energy consumption in the world; the percentage of developing countries will reach 65%.

The digital economy is becoming one of the primary consumers of energy along with industry, transport, and households. It already accounts for 10% of energy consumption. Even super popular mining crypto-currency also requires a large amount of electricity. For example, the Bitcoin alone consumes more than 37 TWh (TeraWatt hours) of energy per year, that is comparable to the annual energy consumption of several European countries. Now even a smartphone that many people use every day has more power than a home refrigerator, if you take into account calls, messages, photo and video transfer, charging, numerous cloud servers, data centers, and radio base station complexes. Indeed, the needs are increasing as the population grows, and its welfare becomes.

However, until now, 68% of the power generation on the planet is produced by thermal power plants (TPP) using fossil fuels. While energy efficiency is becoming increasingly important for public policy, more than 70% of EU coal-fired power plants are about over 20 years old and work with an ability of 32–40%. Technological developments of the 1990s made it possible to increase the efficiency of new TPPs to 40–45%. Real, according to estimates of the European Commission, this is not enough: the highest efficiency of existing technologies is 46–49%. The ability of gas TPPs is on average 52% with a maximum of 58–59%. Gas and oil boiler houses operate with an average efficiency of 36%, and only the introduction of the most modern technologies will allow them very soon to obtain an efficiency of 47%.

The world energy in 10 years will face new challenges. Leaders of several developed countries, frightened by the sociopolitical and environmental risks of possible accidents at nuclear power plants (NPPs), abandoned the development of nuclear energy in favor of heat at the stage of lower oil prices in the 1990s. As a result, the share of nuclear power in the total amount of electricity generation on the planet fell twice—from 19 to 9%. The percentage of TPP is 68%. At the stage of expensive oil, the idea of "green" energy—the use of renewable energy sources (RES), which became in the wake of the Kyoto Protocol and the Paris Climate Agreement idée fixe for many politicians and businessmen, gained the most popularity. Despite the high cost and scientifically proven extreme inefficiency of all alternative energy devices, such as solar panels, wind generators, due to the very low density of energy flow and power, the "green" energy has become almost a religion worldwide.

RES require massive state subsidies; where the renewable energy sector makes up a significant share in the energy balance, for example, in Germany and Denmark, the cost of electricity for consumers is several times higher than where TPP run on coal.

The costs of solar energy, both direct and indirect, are incredibly high. The technology of making photovoltaic panels from an environmental point of view is "dirty." The density of the flow of solar energy is very low. At the same time, the costs of converters, accumulators, regulators, coolers, transformation systems, and transmission of electricity are enormous. And the change of day and night, seasonality and climatic factors make the work of solar installations unstable. The daily operation of a conventional gas turbine or electric generator is much cheaper and more straightforward than a comparable solar power plant.

However, overly optimistic forecasts for renewable energy and electric vehicles, which many governments rely on in their decisions, as well as the fall in oil prices in 2014–2016 and the relative success of the "shale revolution" led to a decline in investment in the oil and gas industry over three years for $750 billion. The OPEC Agreement limits the volume of oil produced. That led to a sharp reduction in exploration and development of deposits. At the same time, the world economy continues to grow at an average rate of 3% per year, and the demand for crude oil in 2018 will be 1.4–1.6 million barrels per day higher than in 2017. That means that already in the coming year, humanity may face a shortage of oil in the market, and no "green" energy can provide growing energy consumption.

Nuclear power has several economic features that distinguish it from the majority of competing technologies. These include:

- High technology intensity, requiring a large amount of research and development work to ensure and justify nuclear radiation safety;
- The enormous volume and more extended period of design and construction;
- Low operating costs and fuel costs;
- Smaller than those of thermal stations and renewable energy resources, the possibilities of working in variable modes.

According to the forecasts of the same scientists, the power of world nuclear energy will increase threefold and by 2050 it will reach the level of 1 million megawatts. That will reduce annual atmospheric emissions of TPP that burn coal to 1.8 billion tons of carbon per year, which will be about 25% of the volume of carbon emissions. In this regard, it is evident that nuclear power has tremendous development potential.

The technological cycle of nuclear power generation includes the production of nuclear fuel for NPPs; energy production at NPPs; exposure of spent nuclear fuel (SNF) at NPPs (up to 3–5 years) in order to reduce the radiation of SNF to a level that allows it to be safely transported in protected containers; transportation of SNF to radiochemical plants; processing of SNF at plants into new nuclear fuel.

A unique feature of the nuclear power industry is that, from an economic point of view, the construction of facilities for the reprocessing of SNF is only beneficial if NPPs produce more than 1.5 tons of SNF per year. This number can produce about 50 large NPPs. However, many countries do not have such a large number of NPPs, so they have to reprocess and store SNF in other countries (Russia, Great Britain, France, China), where there are radiochemical plants for processing SNF. This circumstance leaves its imprint on the nuclear power industry of various countries, giving each country distinctive features.

In Russia, Britain, France, India, and China, NPPs include nuclear fuel production plants for NPPs, NPPs, and radiochemical plants for the reprocessing of SNF, as well as plutonium storage sites and radioactive waste disposal facilities. Also, it includes containers and vehicles for the transport of SNF and radioactive waste.

In the US, there is a slightly different structure of nuclear energy. There, the radiochemical plants have been mothballed, so the SNF from the NPP goes to the storage facilities as a strategic reserve.

Many countries from the entire nuclear power complex have only NPPs. Such states purchase nuclear fuel in the countries that supply nuclear fuel. Some countries recycle SNF at foreign radiochemical plants. For example, Germany, Belgium, Switzerland, Japan, and several other states are processing plants in the UK and France, and Ukraine, Armenia, and Bulgaria—in Russia.

The nuclear weapons complex usually consists of the enterprises of the production cycle for the manufacture of nuclear ammunition and the laboratory and test complex, where they are developing, testing new nuclear ammunition, developing advanced technologies for their manufacture, storage, and disposal. The peculiarity of the complex is that some of its enterprises have a dual purpose and are used simultaneously in the nuclear weapons complex and the nuclear energy sector. Plutonium is produced at the facilities of the first group, which is used as a raw material for the production of nuclear ammunition or fresh fuel for NPPs. Also, the complex includes warehouses for the storage of nuclear ammunition, containers, and their means of transportation.

Determination of the Cost of NPPs

The cost of a conventional thermal power plant is made up of direct and indirect investments, fuel costs, operating costs, and maintenance costs. During the lifetime of a coal-fired power plant, fuel costs make up 50–60% of all charges. For NPPs, investment in construction prevails, accounting for about 70–80% of all costs, while the cost of fuel and other expenses are minimal. Therefore, the value of atomic energy is determined approximately as follows: two-thirds of the cost of energy produced is fixed costs (which will be spent regardless of whether the nuclear power plant will work or not), and the rest falls on production costs.

In the early 1970s, last century, the world economic prospects were very favorable for nuclear power: the demand for electricity generated and the prices of the main types of fuel—coal and oil. As for the cost of building a nuclear power plant, almost all experts were convinced that it would be stable or even begin to decline. However, already in the 1980s, it became clear that these estimates were erroneous: the growth in electricity demand had stopped, the price of fossil fuel not only but even began to decline, and the construction of a nuclear power plant was much more expensive than expected in the most optimistic forecast. As a result, nuclear power engineering everywhere has entered a period of severe economic difficulties. For example, the elimination of all 17 German NPPs that are to be decommissioned before 2022 will cost at least 18 billion euros. Meanwhile, the total amount of spent fuel storage costs was not included in the full amount. Taking into account the cost of storing SNF, the final amount of abandonment of nuclear energy can grow to 30 billion euros. The possible loss of liquidation of one nuclear installation may be from 670 million to 1.2 billion euros.

The undoubted merit of NPPs in the absence of harmful emissions of toxic and greenhouse substances, and in this regard, nuclear power exceeds the sources of energy from fossil fuels and natural gas. Therefore, some countries use this factor as a strategic factor to promote the development of nuclear energy. The accidents with the melting of the core in Chernobyl in 1986 and, especially at the Fukushima-1 nuclear power plant in 2011, forced many countries to pay more attention to the safety issues of NPPs, environmental protection and other aspects of the nuclear fuel cycle. Nuclear power engineering has several interrelated problems, on the solution of which the prospects for its further development and existence in general depend. These include:

a. Issues of safe operation of NPPs;
b. Problems of disposal of radioactive waste and SNF;
c. Risks of the proliferation of nuclear materials.

These problems are technological, and as a result of further development of nuclear technologies can be solved. An additional factor that reduces the large-scale development of nuclear power is the limited availability of uranium reserves. As for the economics of NPPs, it is now challenging to determine all the costs of producing nuclear electricity, including the disposal of radioactive waste and SNF and decommissioning of NPPs. However, the results of estimating the value of atomic electricity, based on all factors, show that it is the same or exceeds the cost of electricity at coal or gas stations.

At present, there is a decline in the nuclear power industry. Its share in global electricity production decreased from 17.6% in 1996 to 10.8% in 2013. The massive two-thirds of the reactors under construction are in China, India, and Russia. Over the past twenty years, the number of orders for new nuclear facilities worldwide has declined. That was due to a whole range of reasons, including public resistance to the construction of new NPPs and the difference in overcapacity of power units in many energy markets, which is mainly due to increased competition in the electricity market and the low economic performance of many existing nuclear facilities. Energy-efficient technologies, which are characterized by low capital costs, the possibility of fast construction and guaranteed operational performance, which is precisely what modern nuclear reactors do not have, win.

Reference

Wallerstein, I. (1974). *The Modern World-System: Capitalist Agriculture and the Origins of the European World-Economy in the Sixteenth Century*. New York: Academic Press.

5

Military Aspects of Nuclear Energy

Nuclear Factor in World Politics: Myths and Reality

Even in 1924, Einstein and the Indian scientist SN Bose studied the behavior of atoms at shallow temperatures; this is known as Bose–Einstein Thermo Dynamics. In Germany in 1939, the German physicists O. Hahn and F. Stresemann discovered the fission of uranium nuclei under the action of neutrons. Less than a month after this great news reached all the leading physical laboratories, already at a conference in Washington, scientists already half-jokingly and half-seriously talked about the release of nuclear energy. However, first, it was necessary to answer the fundamental question: how many new neutrons are produced in each fission event? Are they enough for a chain reaction? Painstaking experiments began, and as a result, the conjectures were confirmed: it was found that the division of one uranium nucleus releases an average of several neutrons—in the order of one, two, or three. So, the prerequisites for the chain reaction were found. Then it turned out that the main component of natural uranium—uranium-238—absorbs neutrons to a greater extent than to divide them. And the central fissile part, uranium-235, is only 0.7% in the natural mixture of isotopes. While the fission of uranium-235 produces enough

D. Basu and V. W. Miroshnik, *The Political Economy of Nuclear Energy*,
https://doi.org/10.1007/978-3-030-27029-2_5

neutrons, most of them disappear in vain, colliding with numerous nuclei of the main isotope. How to deal with this problem?

The first distinct method is the enrichment of uranium ore and the release of pure uranium-235 isotope. But at that time, the separation of uranium isotopes on an industrial scale was too time-consuming, and in other words, almost hopeless. Moreover, at that time no one knew for sure and could not count how many kilograms or tens (and maybe hundreds) of kilograms of this product would be needed for testing.

While the fission of uranium-235 produces enough neutrons, most of them disappear in vain, colliding with numerous nuclei of the main isotope. How to deal with this problem? The first distinct method is the enrichment of uranium ore and the release of pure uranium-235 isotope. But at that time, the separation of uranium isotopes on an industrial scale was too time-consuming, and in other words, almost hopeless. Moreover, at that time, no one knew for sure and could not count how many kilograms or may be tons of this product would be needed to start a chain reaction.

The second way assumed the use of a natural mixture of uranium isotopes and relied on the results of experiments, which indicated that more slowly flying (so-called thermal) neutrons divide the nucleus of uranium-235 more efficiently and are much less absorbed by the nuclei of uranium-238. Consequently, by slowing down the neutrons, one can reduce parasitic absorption and increase the probability of fission. Retarders were also known: light substances such as water, carbon, or beryllium.

Since neutrons are the main participants in the chain reaction, it is not surprising that the experiments on the construction of the first atomic "boiler" were led by the chief neutron specialist and Nobel Prize winner Enrico Fermi, who emigrated from fascist Italy and worked at Columbia University in the United States. A trial installation was started in 1941. Carbon in the form of graphite was chosen as a moderator for a chain reaction.

In 1942, a team of physicists under the leadership of Fermi moved to Chicago, to the Metallurgical Laboratory, where all the leading scientific forces for the study of the chain reaction were concentrated. The experiments continued. Industry under pressure from the military gradually increased the quality of the supplied graphite and enriched uranium, and in the end, judging by the experimental data, it became sufficient for the implementation of a chain reaction.

In 1939, Albert Einstein addressed the US President with a letter in which he justified the need for large-scale experiments to study the possibility of creating an atomic bomb. The scientist's proposal was substantiated by the available information about the successes of Nazi Germany in this field.

Shortly after Einstein's appeal to the president, research began in the United States. After Japan attacked the United States at the end of 1941, these works became more large-scale, and in August 1942, all activities to build an atomic bomb were merged by the "Manhattan Project" in Los Alamos under the leadership of Robert Oppenheimer, and in December 1942. In the United States, the world's first nuclear reactor was launched. Less than three years later, in July 1945, the first atomic bomb was successfully tested in the state of Nevada. The work was conducted in the strictest secrecy. In August 1945, the Americans atomic bombed the Japanese cities of Hiroshima and Nagasaki, demonstrating the reality of the nuclear bomb to the whole world.

Even in the prewar years, the scientists of Moscow, Leningrad, Kharkov and other institutes of the country made fundamental discoveries and calculations on the great chain reaction of uranium fission and the conditions of a nuclear explosion. However, the outbreak of war with Germany with its tragic consequences for the country almost completely stopped the work on nuclear physics. Nevertheless, information about the successes of the American project that came to the USSR through the agents' channels prompted the government to resume its research on the possibility of developing nuclear weapons. For this purpose, in 1943, a special Laboratory No. 2 was created, in which the best scientific and engineering personnel were assembled. The head of the laboratory was appointed I. V. Kurchatov.

In September 1940, the Presidium of the Academy of Sciences of the USSR approved a program of work on the study of uranium fission reactions. Two years later, on September 28, 1942, a secret decree of the State Defense Committee No. 2352 "On *the organization of work on uranium*" was signed, which prescribed to study the feasibility of using atomic energy by splitting uranium core and the possibility of creating a uranium bomb.

After a nuclear strike of the US air forces on Japanese cities Hiroshima and Nagasaki, the work on creating their atomic bomb in the USSR became a national propriety #1. For these purposes, in August 1945, a select interdepartmental committee for solving the nuclear problem for military purposes was created under the leadership of legendary Chairman Lavrenty Pavlovich Beria. In a short time, a specialized nuclear industry was created almost from scratch based on all sectors in the country, ensuring the production of the most diverse and completely new types of specialized equipment, equipment, materials, large-scale construction of nuclear facilities in various regions of the country.

In 1946, for the first time in Europe, a self-sustaining chain reaction of uranium fission was carried out in the F-1 reactor under the direction of Igor Kurchatov, in 1948 the first industrial plutonium production reactor A was launched at secret Company No. 817 (code name "Mayak"). The nuclear industry in Russia was developing in harsh conditions after devastating WW2: a limited material and technical base, knowledge of the biological effects of radiation were limited, the terms of work were dire. Nevertheless, a high level of domestic science and a potent concentration of economic potential made it possible in the shortest possible time to design and launch the first internal nuclear reactor. It happened at the end of December 1946 in Moscow. And on August 29, 1949, the first Russian atomic bomb RDS-1 was successfully tested in the outskirts of city Semipalatinsk.

The path from launching a nuclear reactor to a bomb explosion turned out to be two years and eight months long. With an enormous difference in the material and technical capabilities of the post-war USSR and the United States, Soviet scientists achieved similar results almost at the same time—the first American bomb was detonated after two years and seven months after starting the reactor.

The US abandoned the idea of delivering a nuclear strike against the Soviet Union in the early 1950s; the reason for this was different from what we generally perceive. The United States, as the calculations of the late 1940s showed, had no realizable superiority over the USSR. Atomic warheads of that time were not ways to destroy reinforced concrete structures but were suitable for delivery to targets only by aviation carriers, which could be brought down by fighter aircraft. A full thermonuclear weapon appeared only in the mid-1950s, and it became rocket-delivered nuclear weapon five or seven years later. However, one way or another, by the mid-1960s, the USSR and the United States received the first nuclear weapons in history, which allowed destroying the strategic potential of the opponent in a guaranteed way. The nuclear weapons of other recognized nuclear powers (Britain, France, and China), not to mention the unrecognized nuclear states, do not have this capability to this day (*TASS*, March 26, 2015).

Other functions were assigned to tactical nuclear weapons. Unlike Russian military doctrine, the martial traditions of the United Kingdom and the United States allow striking first using strategic nuclear weapons to compensate for the superiority of the USSR in an army atom. By the middle of the twentieth century, conventional warfare had become too expensive because of the high cost of weapons. Nuclear weapons, like other types of weapons of mass destruction (WMD), were created to make the war cheaper, i.e., removal of the enemy's workforce with less expenditure on weapons systems. The result turned out to be the reverse. The costs of using strategic nuclear weapons blocked the possible benefits of such a campaign. The most likely prices were:

1. The danger of an inevitable retaliatory strike by the enemy (which could be both counterforce and counter-value);
2. Longer-term secondary effects from the use of nuclear weapons.

By the early 1970s, both superpowers were technically able to inflict guaranteed damage to the enemy's strategic potential. On this basis, the concept of "defensive deterrence" was finally established in the United States: to force the enemy to abandon aggression, threatening to inflict unacceptable damage on him. The Soviet leadership did not formally postulate such tasks, but judging from open publications, it saw in strategic nuclear forces a means to destroy the US strategic potential. After the collapse of the Soviet Union, the Russian leadership adopted the logic of defensive deterrence: the strategic nuclear forces are assigned the task of inflicting "predetermined damage" on the enemy for his coercion to peace.

Strategic nuclear weapons can inflict unacceptable damage on the enemy or, at a minimum, increase the costs for possible aggression for him. However, deep nuclear disarmament will mean a decrease in the ability to inflict unacceptable damage on the aggressor (*TASS*, March 26, 2015).

The presence of nuclear weapons by the great powers can cement the state of conflict in their mutual relations. The nature of this phenomenon is approximate as follows: hand-to-hand combat masters will tell that the probability of winning an aggressor with bare hands in a fight with an aggressor's knife is not very high. With skillful use, the sword gives the attacker a significant advantage. However, the aggressor's fear of losing his knife can save an unarmed person in such a struggle. The blade concentrates all the attention of the attacker, which gives the defender a chance to inflict an unexpected series of neutralizing blows.

Similarly, a nuclear weapon state begins to suspect its potential opponents of wanting to take this weapon away from it. Practically any actions of other players are interpreted as insidious attempts to deprive you of a nuclear shield, which, you are sure, is the only indisputable guarantee of your safety. From this point of view, arms control, proposals for mutual inspections, or reduction of stocks of fissile materials appear in the form of insidious attempts by external forces to make you defenseless. And if someone makes such attempts—for example, makes proposals to reduce nuclear arsenals—then it can only be an opponent plotting aggression at the first convenient opportunity. His actions once again confirm the importance of maintaining significant nuclear arsenals.

The circle was thus closed. The presence of nuclear weapons does not only increase your confidence and self-esteem but also raises suspicion toward other players, hence, creating the prerequisites for the escalation of the conflict. These dynamics are well understood and skillfully used, for example, by the military-industrial lobby of major nuclear powers.

A world with a minimum amount of nuclear weapons will not be like today's world minus nuclear weapons. Too many real and potential conflicts must be resolved before moving to deep reductions in nuclear arsenals. However, bearing in mind the above arguments, it can be assumed that the most extensive nuclear military powers, including Russia and the United States, would be more profitable to stand in the vanguard of the process of nuclear disarmament than to insist on their exclusive right to possess a large number of nuclear weapons. That would significantly increase their credibility in the eyes of developing nuclear-free states.

Perhaps peaceful nuclear energy has no alternatives if humanity seriously wanted to reduce hydrocarbon emissions into the atmosphere. However, military nuclear technology has not saved anyone from anything and has not created an advantage that substantially outweighs the risks.

> "Is it possible to establish a cause-and-effect relationship between the absence of world war during the last seven decades and the existence of nuclear weapons by the world's major powers?" It is impossible to answer this question unequivocally, as, however, on most philosophical issues. However, the author ignores the fact that nuclear weapons, like any weapon, have a clear purpose and patterns of application powers. (*Interfax*, December 19, 2012)

In the case of total nuclear war of human civilization, as we know it, the end is likely to come. But what happens if all the superpowers launch their entire nuclear arsenal at once on a hypothetical enemy? The explosive power of a nuclear weapon is usually measured in kilotons of trinitrotoluene (TNT). The power of the bomb dropped on Hiroshima is estimated at 16,000 kilotons of TNT. The power of the bomb dropped on Nagasaki is estimated at 22,000 kilotons of TNT. The most powerful nuclear weapon ever detonated on Earth in the test site, the "Tsar-bomba," developed in Soviet Russia yielded 50 million tons of TNT.

Since the tragic events—the nuclear bombings of Hiroshima and Nagasaki—more than 70 years have passed. These days it is essential to once again reflect on the role of nuclear weapons in the modern world, on the impact of nuclear arsenals on the security of individual states and terrestrial civilization in general. It would seem that the enormous human sacrifices and suffering that resulted from the bombings in August 1945 underscore the danger of possessing nuclear weapons. The realization of this danger would have to increase as the destructive power, and long-term consequences of the use of a nuclear bomb became evident.

First, the risks of maintaining significant arsenals of nuclear weapons (and raw materials for its production in the form of fissile materials) are now growing as a result of the desire of new terrorist groups to obtain and use WMD.

Secondly, when nuclear arsenals are ready for immediate use, the price of a technical malfunction or human error increases astronomically. Since the beginning of the nuclear era, several such cases have been recorded—for example, the unintentional discharge of a nuclear warhead from a bomber during a training exercise in the United States or the erroneous loading of missiles with nuclear warheads into an airplane flying several thousand kilometers over American territory.

In all these cases, it was possible to avoid the explosion of a nuclear bomb or a massive launch of nuclear missiles only by a fluke or due to the extraordinary actions of personnel. Finally, even if the technical systems work well, and the security of nuclear materials is ensured at a high level, it is quite possible that the leaders of one nuclear power could misinterpret the intentions of another nuclear power, especially in the conditions of the "fog of war"—the enormous nervous tension and uncertainty inherent in any conflict.

Despite these risks and threats, popular opinion among politicians and experts remains the opinion on the "stabilizing role" of nuclear weapons. Supporters of this idea argue, among other things, that the cold war did not develop into a hot war, mainly due to the mutual nuclear deterrence of superpowers. However, in practice, to prove that nuclear weapons can prevent conflict without creating unjustified risks is not so simple (*TASS*, March 26, 2015).

First, is it possible to establish a causal relationship between the absence of world war during the last seven decades and the existence of nuclear weapons by the world's major powers? In the past 70 years, has it been these weapons that stopped those politicians who were considering options for an aggressive war against the opposing forces?

Even if in some cases it was the nuclear factor that played the central deterrent role, the parties came too close to the brink of nuclear war so that the outcome of the conflict could be called the success of nuclear deterrence. For example, did the risk of nuclear war compel the USSR and the United States to stop escalating during the Caribbean crisis of 1962? Perhaps the US president Kennedy and First Secretary of the Central Committee of the USSR Nikita Khrushchev stopped a step away from a direct clash precisely because of fear of nuclear war. However, such a happy outcome cannot be considered predetermined. The right approach to the "red line" sharply increased the likelihood of exchange of nuclear strikes due to the "fog of war" or technical malfunction. Therefore, the relatively successful outcome of the Caribbean crisis can hardly be considered a confirmation of the effectiveness of nuclear deterrence.

Also, some countries that possess nuclear weapons or the ability to quickly develop them have an increased sense of vulnerability, which creates significant risks of conflict with the use of nuclear weapons (as scientifically proven, any such dispute would have catastrophic global consequences). For example, Pakistan has repeatedly openly declared its readiness to use nuclear weapons even in response to actions that do not contain signs of classic armed aggression—that is, a direct military invasion of Pakistani territory by India.

Several other circumstances also raise doubts about the fundamental effectiveness of nuclear deterrence. Thus, nuclear weapons in the hands of London did not save the British Empire from a final collapse, and Paris was not allowed to prevent the separation of Algeria (where the first French nuclear test site was located). In 1973, a coalition of Arab states attacked Israel, although at that time there was reliable information about Israel's nuclear deterrence potential. The enormous nuclear arsenals did not prevent the USSR from disintegration.

A list of similar examples can be continued. It is more important, however, to understand that the possession of nuclear weapons often has psychological consequences, the reverse restraints of potential opponents and the self-containment of the nuclear state.

Why Do We Need Nuclear Arms Control?

The task of Soviet-American, and later Russian-American, arms control agreements were not nuclear disarmament, but a mutual reduction of counterforce capabilities.

The USSR/Russia and US arms control agreements pursued three goals:

1. The introduction of limits on the "destabilizing" components of the strategic triad, for example, the dividing heads of individual guidance;
2. Elimination of nuclear weapons carriers, which can be used for counterforce or counter-elite strikes (medium- and shorter-range ballistic missiles);
3. Coordinated elimination of obsolete warheads and their carriers.

None of these tasks challenged the logic of strategic parity. Doubts emerged in the 1990s, which was due to the development of non-nuclear precision weapons and anti-missile defense (ABM) systems. Against this background, arms control began to lose its stabilizing role. The Russian side has faced a dangerous trend: a small number of nuclear weapons may not survive the disarming strike of the enemy and to stop operations of his missile defense systems; or, if he survives, will not be able to inflict unacceptable damage on the enemy (*TASS*, March 26, 2015).

That type of discussion ceased in the 1970s, which was due to the development of non-nuclear precision weapons. Today, when modeling the "optimal" number of strategic nuclear forces, one must take into account not only the counterforce capabilities of the enemy but also its non-nuclear weapons and missile defense systems.

Nuclear Weapons and the Survival of the Globe

The use of nuclear weapons for the sudden destruction of the US's nuclear potential (or Russia's, as well as virtually any modern nuclear power) is unthinkable since in any case, it will lead to a global environmental catastrophe. The scale of such a calamity was not completely known to the scientists, who in the 1970s were worried about the threat of a sudden disarming nuclear strike by one superpower over another. It may be recalled that a decade earlier, in the 1960s, nuclear strategists in the United States and the USSR seriously discussed the prospects of life after the exchange of nuclear strikes, which would have destroyed less than a quarter of the population and half of the industry of the world. It was not until the mid-1980s that scientists proved that a sudden massive nuclear strike (not to mention the nuclear war) would become suicidal for its initiator and therefore, in principle, could not be successful.

High-precision non-nuclear weapons, actively developed today, at least by the USA, Russia, and China, represent a problem as a means of rapidly destroying nuclear potential. The threat of depreciation of nuclear deterrence potentials may increase if missile defense systems developed not only by the United States but also by Russia and China, are brought to a high degree of perfection.

However, the presence of this problem rather refutes the thesis about the effectiveness of nuclear deterrence and certainly does not confirm it. An interesting (as yet academic) hypothesis is that the introduction of new types of weapons into the arsenals of the major powers will eventually lead to the dying off of nuclear armaments as inferior to them in the effectiveness of solving urgent problems. If the development of systems of high-precision strategic non-nuclear weapons, missile defense or any other new weapons cannot be stopped (which cannot be ruled out), then will this lead to the "withering away" of nuclear weapons as the exclusive and most important means of protection? Will not it be recognized as more destabilizing due to the irreversibility of the destructive consequences of its application? Could it not be sufficient for strategic containment of a weapon capable of inflicting commensurate damage without the risk of a "nuclear winter" for the whole of humanity?

One believes that by signing nuclear arms control agreements, the USSR/Russia and the United States merely imposed "limits on the" destabilizing "components of the strategic triad, eliminated certain nuclear weapons carriers and paved the way for concerted elimination of obsolete warheads and their carriers." These considerations were undoubtedly crucial for both sides in terms of maintaining bilateral nuclear parity. More importantly, however, Moscow and Washington publicly presented their agreements as a sign of the two countries' conscientious desire for general and complete nuclear disarmament in fulfillment of their obligations under Article VI of the Treaty on the Non-Proliferation of Nuclear Weapons (NPT), which is now defunct as intended by the US President Trump. Both the USSR/Russia and the US have traditionally been concerned about compliance with their obligations under the NPT and sought to minimize the possibility of criticism for Moscow and Washington from non-nuclear states for lack of adherence to the NPT (*TASS*, March 26, 2015).

One can reasonably believe that the nuclear shield should not be "risked." It was suggested only to stop attributing to nuclear weapons those properties that it does not possess—first of all, the ability to protect the state from any threats other than direct and large-scale (and therefore unlikely) military aggression.

In turn, the risks that these weapons generate for international security are quite real. One can correctly note that terrorist groups, fortunately, failed to obtain nuclear materials during the previous twenty or more years. However, the majority of the developed countries of the world, including the United States and Russia, recognized the fact of an increase in the transnational terrorist threat, as well as an increase in the financial and organizational capabilities of real and potential terrorists.

Are Nuclear Weapons Useful?

In the 1930s, a notorious group of Soviet and French military leaders opposed the "reassessment of the role of tank units." In 1956 Khrushchev said that aviation and the surface fleet were outdated in the era of missile weapons. In each of these cases, the payment for underestimating one or another type of weapons was painful. As early as the 1820s, the eminent German strategist Karl von Clausewitz derived the formula: "*War is the continuation of politics by other means.*" The armed forces, including nuclear weapons (NW), are not an end in themselves, but a tool for solving political problems. Their goal is to achieve a military victory, which politicians can (or cannot) capitalize in a political outcome.

By the way, neither Britain nor France technically could resort to nuclear blackmail of the USSR or the United States even immediately after the creation of nuclear weapons—given the absence at that time of delivery vehicles of nuclear warheads to the territories of their opponents. Gorbachev asserted that Clausewitz's formula on the correlation of war and politics ceased to function in the nuclear age. Nuclear weapons, in his opinion, could not be considered a means of achieving a military victory as there cannot be a winner in a nuclear war. This thesis enabled Gorbachev to declare the priority of universal human interests over class interests, above all, the idea of "*the global survival of mankind in the nuclear age.*" Gorbachev's acceptance of several international self-restrictions was also carried out on his initiative. The administration of President Ronald Reagan, by the way, was very critical of Gorbachev's efforts on general nuclear disarmament and tried to get away from them. Nuclear weapons have two tasks: (1) Military: to defeat the strategic potential of the opponent by inflicting a counter-value, counterforce, or counter-elite ("decapitation") strike, and (2) Military-political: to prevent military aggression against a nuclear power by threatening unacceptable damage to the enemy; officially, this task was postulated by the USA, Russia, France, and, in part, Pakistan. Not a single nuclear power has put formally other functions like maintaining international stability before atomic weapons.

Before the Second World War, the great powers actively developed bomber aviation, designed to destroy enemy aircraft at airfields. Did this mean that the USSR, for example, had to recognize the destabilizing role of its aviation or its low utility? Similarly, in the 1950s, both the USSR and the United States actively developed air defense systems (anti-aircraft defense) to defend against strategic enemy aircraft. The air defense complexes, of course, became the primary targets for the armed forces of the opposite side. Did that mean the USSR or the United States should have recognized the destabilizing role of its air defense and dismantle it as a tempting opponent?

The war, as the researchers of the nineteenth century proved, is caused by political, and not military, reasons. The state of the armed forces can encourage the aggressor to commit aggression, or, conversely, dissuade him from doing so. But the armed forces themselves (including their nuclear component) do not start military operations and do not provoke them. In other words: reproach for destabilization does not need nuclear weapons, but politicians who want to destroy nuclear capabilities by a preemptive strike.

The American nuclear strategy was initially oriented toward the defeat of a strategic opponent (primarily the USSR). At the first stage until 1961, it provided for a counter-value strike: the failure of cities and industrial targets of the enemy. After 1961, priority was shifted to counter-striking: the destruction of the enemy's strategic nuclear forces before their activation.

The Americans did not reject the concept of a counter-value strike. The administration of Bill Clinton in 1994 adopted the idea of the possibility of "selective defeat of a set of goals" of nuclear weapons for compelling the enemy to peace as a combination of counterforce and counter-value strikes. Any version of the "National Military Strategy" postulates the right and the ability of the United States to deliver a preemptive nuclear strike against opponents.

British nuclear strategy has developed the theory and practice of using tactical nuclear weapons. At the turn of 1945–1946, British military strategists have pondered over whether it is possible to replace a powerful land army with the help of nuclear weapons. By 1953, before the creation of its nuclear weapons, the "nuclear menu" of the Kingdom had already singled out three types of nuclear strikes:

1. Demonstration (demonstration of readiness to use nuclear weapons in the event of a continuation of the war);
2. Operational-tactical (defeat of enemy objects in the nearest strategic rear);
3. Defensive (the destruction of groups of the armed forces of the enemy in the theater of military operations).

By about 1957, British developments became the basis of the US strategy of using tactical nuclear weapons. The French nuclear policy is officially aimed at inflicting a counter-value attack on key political targets of the enemy. This task is confirmed by the latest version of the White Paper on Defense Issues of 2013. The Soviet military doctrine allowed the use of nuclear weapons in cooperation with other branches of the armed forces. Russia in 1997 adopted the American concept of deterrence, and in 2000 abandoned the Soviet commitment not to deliver the first nuclear strike. Since then, the Russian military doctrine is oriented to inflicting on the potential aggressor a "predetermined damage."

Objectives for the use of nuclear weapons have long been established and are continually being improved. While there is no political solution for their defeat, but this does not mean the absence of goals as such. To date, only nuclear missile weapons can be guaranteed to hit strategic enemy targets. Non-nuclear aviation, as it was recognized back in the mid-1940s, is not able to solve such a problem because of countering the air defense forces and fighter aviation. Non-nuclear precision-guided weapons are also incapable of solving such a problem due to the insufficient striking power of high-precision systems. At best, they are still suitable for defeating the infrastructure of weak countries. Controversial and proposed by the author radical reduction of nuclear weapons: the nuclear powers cannot technically take hostages to a strategic arsenal of the opponent. The fact of the matter is that the SALT-1 (1972) and SALT-2 (1979) contractual complexes assumed not the conquest, but the agreed arming of the USSR and the United States up to certain limits. Both superpowers needed more nuclear weapons to take a hostage to the growing strategic potential of the opponent. The situation changed by the mid-1980s when the equipping of nuclear forces with the separable warheads of individual targeting allowed to solve the same problems with a smaller grouping of carriers. That's when it became possible to proceed to an agreed elimination of carriers. In a world with a small number of nuclear weapons, Russia and the United States will be forced to turn to the French atomic strategy of counter-value strikes. Unable to hit strategic potential, they will have to reorient strategic nuclear forces to defeat crucial enemy targets. Such effects will be more natural to destroy by counterforce strikes.

Approximately 93% of all nuclear warheads are owned by Russia and the United States who each have around 4000 warheads in their military stockpiles; no other nuclear-armed state sees a need for more than a few hundred nuclear weapons for national security.

> As of early 2018, we estimate that Russia has a stockpile of roughly 4,350 nuclear warheads assigned for use by long-range strategic launchers and shorter-range tactical nuclear forces. Of these, roughly 1,600 strategic warheads are deployed on ballistic missiles and at heavy bomber bases, while another 920 strategic warheads are in storage along with about 1,830 non-strategic warheads. In addition to the military stockpile for operational forces, a large number – perhaps almost 2,500 – of retired but still largely intact warheads await dismantlement, for a total inventory of more than 6,850 warheads. The United States, Russia, and the United Kingdom are reducing their warhead inventories, but the pace of reduction is slowing compared with the past 25 years. France and Israel have relatively stable

inventories, while China, Pakistan, India, and North Korea are increasing their warhead inventories.

Reference

Kristensen, H. K., & Norris, R. (2018). Russian Nuclear Forces, 2018. *Bulletin of the Atomic Scientists, 74*(3), 185–195.

Part II

Retrospect of Nuclear Power Development

6

History of the Development of Nuclear Power

The history of the most common notions about atoms, the smallest particles that form matter, is usually traced back to the time of the ancient Indian philosophers Uddalak Aruni (8th century BC), Konad (6th century BC) (Dasgupta 1922; Radhakrishnan 1923), and the ancient Greek philosophers Leucippus and Democritus (460–370 BC). They represent the founders of an atomic theory of the universe and materialistic philosophy. The Greek philosopher Epicurus (341–270 BC) adopted this nuclear approach. In the 1st century BC one of his followers, the famous Roman poet and philosopher, Lucretius Car (99–55 BC) outlined the teachings of Epicurus in his poem *On the Nature of Things* through which they were preserved for future generations (Brodov 1984; Freeman 1948).

However, one of the greatest scholars of antiquity, Aristotle (384–322 BC) rejected atomistic theory. Unfortunately, his views on philosophy and science later prevailed in the scientific thinking of the Middle Ages. Thus, for example, on September 4, 1626 the Parliament of Paris in France banned the teaching of atoms on pain of death. Could the great Democritus have imagined that over 2000 years later in Europe the state would execute followers of his atomic theory? The Parliament of Paris in France was not the first to ban atomic theory. Initially, the theory was rejected

D. Basu and V. W. Miroshnik, *The Political Economy of Nuclear Energy*,
https://doi.org/10.1007/978-3-030-27029-2_6

1200 years earlier by Dionysius of Alexandria, St. Augustine, and other Church fathers. The Church severely condemned the doctrine that all bodies of nature were formed without the help of divine forces but instead by the unification and separation of invisibly small and indivisible particles—atoms. So, after the collapse of the ancient world the Church completely seized and promoted spiritual life—the consequence being that atomic theory for thousands of years was no longer used by scientists.

During the Renaissance research began on what is now known today as chemistry and physics bringing new insights into the nature of the mysterious *indivisible particles* called atoms. Although Robert Boyle (1627–1691) and Isaac Newton (1643–1727) proceeded in their reasoning from the idea of the existence of indivisible particles of matter, neither needed a detailed atomistic theory to explain the phenomena that interested them. Therefore, the results of their experiments did not add anything new about the properties of atoms. The atomistic theory had all but been forgotten up to the end of the Renaissance. Then in the eighteenth century the great Russian scientist Mikhailo Lomonosov (1711–1765) experimentally proved that the doctrine of atoms and molecules was the real foundation of physics and chemistry on which it was possible to build all the most important laws of these sciences without resorting to mysterious "divine forces."

Lomonosov's works allowed the prediction of unknown phenomena for the first time. He turned atomic theory into a reliable tool for scientific research. In the nineteenth century another great Russian scientist Dmitry Mendeleev (1834–1907) discovered the fundamental law of atoms. This law states that the atoms of various elements differ in their properties but not by chance. Mendeleev predicted the natural sequence of items. Such a law of the logical sequence of features he called a periodic law and the entire system a regular system of elements. This is now known as Mendeleev's periodic system of details or Mendeleev's Periodic Table. Mendeleev's law regarding the periodic repetition of the properties of atoms in the natural sequence of elements is the fundamental law of particles and one of the essential requirements of nature. It follows from the periodic law that if there is an element with the properties of a pronounced metal, there must necessarily exist another one with less pronounced metal properties followed by a component with even weaker metal properties, etc. Moreover, if some of them are missing, it is not because they do not exist in nature but because they had not yet been discovered. Where the natural sequence of atoms was disturbed by the absence of such undiscovered elements, Mendeleev left empty spaces in his periodic system of elements. He was absolutely sure these elements existed and sooner or later would be discovered. To facilitate their discovery Mendeleev predicted their most relevant properties. In this way he offered scientists the opportunity to test his law.

But where did the genius Mendeleev get his atomic weights? For example, why did he give uranium the atomic weight of 240 instead of the "120" believed by all other scientists? His answer was because uranium's nuclear weight had been miscalculated. For the first time the periodic law made it possible to judge what the atomic mass of an element should be. After all, giving uranium the nuclear weight of "120" in the natural sequence of atoms meant it would have to be placed in the space occupied by an element whose properties had nothing to do with uranium.

On the other hand, the atomic weight of 240 placed uranium in the same group of related elements regarding their characteristics. This reasoning led Mendeleev to correct the atomic weights of other items in his periodic system of elements. All his corrections were soon empirically confirmed. Therefore, the creation of Mendeleev's regular system of elements brought vagueness in chemistry to an end. Mendeleev's periodic system of components and his periodic law—the fundamental law of atoms—was considered by scientists a bright guiding star that has ever since pointed researchers in the right direction. Uranium was the last element discovered by chance. Discovered in 1789 by Martin Klaproth (1743–1817), a prominent German chemist, uranium was named for the planet Uranus. Later on all discoveries of new elements were made based on clear guidelines provided by Mendeleev's periodic law.

Most prominent scientists supported Mendeleev's theory that states the elements are arranged in a natural sequence positioned in ascending order of their atomic weights. For example, in 1880 the famous Swedish scientist Lars Nielson (1849–1899) discovered the second element predicted by Mendeleev and named it scandium in honor of Scandinavia. This prompted him to write: "This is how the thoughts of the Russian chemist are confirmed, which allowed not only to foresee the existence of this simple element, but also to provide it with its most important properties". In 1885 the prominent German chemist Winkler (1838–1904) discovered the third element predicted by Mendeleev and called it germanium. This too prompted him to write to Mendeleev: "I inform you of the new triumph of your ingenious research and show you my respect and deep respect".

The ability to predict the discovery of elements was the first consequence of applying Mendeleev's periodic law. This law not only reawakened researchers' thoughts but brought them to the fore. Why do the properties of atoms change according to rules put forward by Mendeleev's periodic system of elements? What does it say about the inner complexity of atoms? The famous Russian chemist Alexander Butlerov, a friend of Mendeleev, was the first to pose such a question in the history of science. He wrote in 1886 in his book *Basic Concepts of Chemistry* that "atoms are not indivisible in nature, but are indivisible only by the means now available to us and… can be divided into new processes that will be discovered later".

That bold idea of Butlerov attracted little sympathy from the academic community at first. However, new facts began to accumulate that soon confirmed his idea and his brilliance. It was noticed, for example, that any element (especially, metals) are positively charged when they are incandescent and when they are illuminated with ultraviolet light. Since they were previously electrically neutral they lose negative electrical charge when exposed to these effects. Moreover, since it was known that matter consists of atoms and emptiness it had to be the particles themselves that were losing negative electricity.

This discovery, in fact, means that atoms that had been considered for more than 2000 years totally unbreakable are divided into at least two parts: negative electricity and positively charged residues. It was soon established that the negative power included in atoms was the same as electricity that flows through the wires of an electrical network. Its particles have the same dimensions, weight, and charge. These tiny particles of electricity—atoms of electricity—were given the name electrons. The discovery of the electrical features of atoms made their study extremely easy. Subtle and precise electrical methods helped rough measurement and weighing giving rise to the long-awaited opportunity to determine the size and weight of very tiny particles.

Thus, when discussing the history of nuclear science it is essential to start the analysis from the beginning of the twentieth century. At that time the neutron had not yet been discovered, there was no acceptable theory of the structure of the atom, and the possibility of a chain reaction of nuclear fission had never been discussed.

By 1900 the outstanding Russian scientist Vladimir Vernadsky (1863–1945) wrote a report for the Russian Academy of Sciences on the development of new nuclear forces based on radioactivity research into radium. At that time the Nobel Prize winners Marie Skłodowska-Curie and Pierre Curie were developing their well-known experiments on radium. Vernadsky understood that nuclear forces existed that were a million times more effective than currently known chemical forces. He suggested that humanity was entering a new atomic era in which there would be no restrictions on access to energy and that all problems of nutrition, health, and industrial and social development would be solved. That was one of the first predictions made about the coming golden age of humanity based on scientific and technological revolution. Later, however, it dawned on him that by opening the door to exploration of atomic energy it might lead humanity not only to a brighter future but also to the possibility of self-destruction.

European Research Schools on Nuclear Physics

One of the first experimental–theoretical schools to play a historical role in the development of nuclear energy was the Ernest Rutherford (1871–1937) School in the United Kingdom. Scientists had always assumed that individual atoms were indivisible and unchangeable. However, British physicist Ernest Rutherford having studied the penetrative power of uranium radiation showed that there were two types of radiation: very "soft" radiation easily absorbed by a substance that Rutherford called alpha rays, and more penetrative radiation that he termed beta rays. The beta rays turned out to be identical to conventional electrons.

Alpha rays were found to have the same charge and mass as helium atoms devoid of their two electrons. The third type of radiation termed gamma rays turned out to be similar to X-rays, but had even higher penetrative capacity. In 1902 Rutherford with the help of his talented young assistant Frederick Soddy (1877–1956) was able to show that when an atom emits alpha or beta rays it is being converted into an atom of another kind. At first, scientists could not believe it. However, Rutherford and Soddy conducted a series of experiments with radioactive decay and transformed uranium into lead. This stunning series of discoveries won Rutherford the Nobel Prize in 1908 (the Nobel Prize was later awarded to his assistant Soddy), but his greatest achievement was still ahead. He noticed that fast-moving alpha particles could pass through thin gold foil (despite being slightly deflected they left no visible traces!). There was an assumption that the gold atoms despite being believed to be stable, impermeable "tiny billiard balls" were soft on the inside! Rutherford's experiments proved that smaller alpha particles could pass through gold atoms like a high-speed bullet through jelly.

In 1911 Rutherford and his student Heinrich Moseley (1887–1915) discovered that positive charges are grouped in the very center of an atom in one single particle that he termed the atomic nucleus and that negative electrons rotate around this nucleus. The speed of rotation of electrons is so high that they seem to create a continuous barrier around the nucleus like a continuous shell. This is the reason an atom under normal conditions behaves like a solid particle. Rutherford also determined the size of the atomic nucleus. It turned out that nuclei are not much larger than electrons: even the nuclei of the heaviest atoms are only 10 times larger than the electron. Such a small difference amazed Rutherford and other scientists. After all, the charge of the nucleus has to balance all the electrons rotating around it. If the charge of the electron is taken as a unit of atomic charge, then the charge of the nucleus is equal to the number of electrons surrounding it. At the same time this multi-charged particle is almost as small as any of these electrons! Thus, the works of Rutherford and Moseley immensely strengthened the value of the periodic law of Mendeleev. In particular, they answered the ongoing research question as to just how many elements and varieties of atoms there were. Atoms differ in their serial numbers (in other words, in the charges of atomic nuclei).

The smallest charge has an atom of hydrogen as "1"; the most significant charge has an atom of uranium as "92." That is the reason Mendeleev placed hydrogen in the 1st cell and uranium in the 92nd cell (the last cell) of his periodic system of elements. There can only be 90 other elements between hydrogen and uranium—no more, no less. Rutherford's research proved the predictions of Mendeleev.

At the same time the Copenhagen School led by the Danish scientist Niels Bohr (1885–1962) began to develop in Central Europe. Together with the Rutherford School it laid the foundations for modern nuclear physics. In 1913 Bohr began to investigate another problem: how are electrons located around the atomic nucleus? The one thing that was clear was that the location of electrons swarming around the nucleus could not be accidental. Analogous to constructing a building with a simple bunch of beams, bricks, glass, and iron, it was necessary to know and take into account all the properties of the building materials that make up a reliable, stable structure. Bohr felt the same logic must apply to the formation of the atom. At that time there was no way of seeing how the nucleus and electrons could create the structure of the atom. However, Bohr found a way: he took into account all the properties of each part of the atom and the mutual influence they have on each other to imagine a picture in which each section would fit so that the whole structure would be the most stable possible, and then check whether this picture explains the actual properties of matter. Lomonosov proceeded in a similar way when creating his theory of the gas state, the theory of heat, and his other theories. Step by step Bohr moved along Mendeleev's periodic table covering sheets of paper with dozens of formulas and thousands of numbers behind which was hidden an elaborate game of forces that determine the conditions for the existence of invisibly small, but exceptionally complex atomic structures. Bohr thought these very tiny structures were so reminiscent of the solar system with the nucleus as a Sun in the center and a swarm of electrons as planets rushing around it. Calculations demonstrated that up to the 10th element of the periodic system—in the atoms of carbon, nitrogen, oxygen, chlorine, and neon—new electrons could be added to the second layer perfectly explaining all their properties.

So, little by little Bohr revealed the structures of the electronic shells of atoms. On every occasion he was able to provide a complete explanation of the properties of the element in question and demonstrate the striking coincidence with the requirements of Mendeleev's periodic law. There was a suggestion that maybe he just adjusted his calculations to these requirements? The only way to confirm the correctness of calculations was by predicting something new and unforeseen without knowing the structure of the electronic shells of that atom. In 1922, after nine years of research, an opportunity cropped up for Neils Bohr to check the correctness of his calculations. At that time the element with the serial number "72" had not yet been discovered. This element was considered to have properties close to the element lanthanum ("57"). Lanthanum ores were searched to see if there was any trace of this element, but all to no avail. However, having calculated the possible location of the 72 electrons in an atom of this undiscovered element, Bohr found that based on its electronic structure and thus its properties element number "72" must be similar to zircon ("40")—not to lanthanum ("57"). Therefore, it was necessary to look for it in zirconium minerals rather than lanthanum ores. Having made this bold prediction Bohr commissioned two of his employees, the Hungarian scientist Heveshi and the Dutch scientist Koster, working in his laboratory in Copenhagen at the time to test his predictions. The following year (1923) Bohr celebrated a brilliant victory: the unknown element number "72" was found in Norwegian zircon ore precisely according to his theoretical predictions. In honor of Copenhagen (*Gafnia* in Latin) the new element was named hafnium.

The discovery confirmed Bohr's predictions based on Mendeleev's periodic law conclusions. Thus, Bohr's research raised the already high status of Mendeleev's law and confirmed the sheer brilliance of the Russian scientist who not only discovered the fundamental law of atoms, but did so with extraordinary completeness and accuracy when it came to reflecting their main properties. In the half-century or so since the discovery of this law the doctrine of atoms has advanced immeasurably more than during the two and a half millennia since atoms were first believed to exist.

Mendeleev's periodic law led to the discovery of the complexity of atoms, the development of their components, and the requirements of their structure. Mendeleev's great law showed scientists all over the world how to discover the mysteries of the atomic nucleus and gain mastery of nuclear energy. In August 1871 Mendeleev wrote in his famous article published in 1872 in a German chemical journal: "The weight is naturally attributed to the first movement of matter, and there is no reason to deny the possibility of transformation of this movement in the formation of atoms in the chemical energy or another type of action" (https://dmitrimendeleev.com/).

In these words uttered in the nineteenth century Mendeleev directly indicated how to master atomic energy. This involved studying the processes of decomposition and formation of atoms, the accompanying change in their weight, and the release of energy. Thus, all research that has taken place anywhere in the world in the mastery of atomic energy is the result of implementing the brilliant predictions of the Russian scientists Mikhailo Lomonosov and Dmitry Mendeleev.

In Germany in 1895 the famous scientist Wilhelm Röntgen (1845–1923) discovered ionizing radiation by passing an electric current through a glass tube and producing X-rays. The ionizing radiation discovered by Röntgen marked an entire scientific era and revolutionized medical diagnosis. As a consequence he became the world's first Nobel Prize winner in physics in 1901.

Later (in 1896) the married couple Pierre Curie (1859–1906) and Marie Skłodowska–Curie (1867–1934) called the phenomenon discovered by Röntgen *radioactivity*. The Curies were the first physicists to study the radioactivity of elements. In 1898 Pierre and Marie found an entirely new radioactive element and named it polonium after Poland, the birthplace of Marie. This soft silver–white metal filled the 86th cell in Mendeleev's periodic table of chemical components. At the end of 1898 the Curies discovered radium, a shiny alkaline earth metal with radioactive properties. It occupied the 88th cell of Mendeleev's periodic table. After radium and polonium Marie and Pierre Curie went on to discover several other radioactive elements. Scientists have found that all the heavy elements located in the lower cells of the Mendeleev table have radioactive properties. In 1906 Pierre and Marie discovered that elements contained in the cells of all living beings on Earth, such as the potassium isotope, possessed radioactivity. These scientists won the Nobel Prize in physics for their contribution to science.

In 1910, after the death of her beloved husband and coauthor, Marie Curie succeeded in collaboration with Andre-Louis Debierne

(1974–1949) to produce pure metallic radium with strange, almost magical properties. She called these properties radioactivity. Thus, a 12-year cycle of research came to an end in which radium was undeniably proven to be an independent chemical element. After this work Marie Curie was nominated for election to the French Academy of Sciences. However, this resulted in a disgraceful scandal when most of the conservative academics determined not to let a woman into their ranks. As a result Marie Curie's candidacy was rejected by a single vote. This decision became all the more shameful when Marie Curie won her second Nobel Prize in 1911, this time in chemistry. Marie Curie was the first scientist to receive two Nobel Prizes. Her brilliance shone through to her family with her daughter and son-in-law also getting the Nobel Prize in chemistry.

In 1935 the daughter of Pierre and Marie Curie, Irène Joliot-Curie (1897–1956) and her husband Frédéric Joliot (1900–1958) were jointly awarded the Nobel Prize in chemistry for synthesizing new radioactive elements. Marie Skłodowska-Curie missed witnessing this triumph of her daughter and son-in-law by just a year; she died in 1934 as a result of a severe blood disease caused by prolonged exposure to radiation. For many years Marie Curie did not receive any financial support for her research and did not even have her own laboratory until 1904 when she had already gained full recognition and fame in the scientific world. To get about 0.1 g of pure radium she needed to process more than 500 kg of uraninite. Decades after Maria Skłodowska-Curie's death, her belongings, scientific notes, and diaries are kept under special conditions and are not accessible to visitors since the level of radioactivity they have is still dangerous for health. Few contemporaries of the Curies understood the importance of their scientific discoveries. Today, the words "radioactivity" and "radiation" are known to everyone. Who hasn't heard of radiation leaks in nuclear power plants and that some cancers and other diseases are treated by radioactivity? However, a hundred years ago no one knew these words. They were invented by Marie Curie and her husband Pierre to describe the ability of some chemical elements to emit elementary particles. Without their work, today there would be no radiation treatment for cancer, no nuclear power, and no new scientific evidence of the origin of the universe. Pierre and Marie Curie are responsible for a revolution in the life of humanity as a result of people learning to harness atomic energy for medical and peaceful purposes.

Research School on Nuclear Physics in the United States

Just before the outbreak of the Second World War a number of Nobel laureates in chemistry and physics led by the famous German scientist Albert Einstein emigrated to the United States. In 1939 two of them Leo Scyllard and Eugene Wigner (both Hungarian physicists) prompted Albert Einstein to write a letter to US President Roosevelt and explain the need to develop an atomic project in America to counteract developments in Nazi Germany. The scientific community understood that nuclear forces could be a source of great evil and advocated the need to control this technology. President Roosevelt took this message very seriously and had a hand a few months later in setting up the Uranium Committee in the United States that played a major part in the Manhattan Project.

Creation of a Nuclear Reactor

In 1938 Otto Gan and Fritz Straussman discovered that uranium nuclei form other elements when their neutrons are being bombarded. Thus, for a nuclear reactor to operate a fissile isotope is needed. In nature there is only one fissile isotope: uranium-235. When a neutron penetrates the nucleus and gets absorbed the nucleus of uranium-235 begins to deform, activated by the binding energy applied of the neutron. As a result the nucleus can be divided into two fission fragments that emit new fission neutrons. When fission neutrons enter the next nucleus it can also split up releasing free neutrons and resulting in a fission chain reaction. If the number of free fission neutrons is controlled by absorbing excess neutrons the chain reaction can proceed normally. The kinetic energy of the fission fragments energizes and heats the surrounding matter. As a result of the nuclear fission reaction the energy released in the reactor is many times greater than the energy of chemical reactions: fission of 1 g of uranium-235 is equivalent to burning 3 tons of coal. When all the fission neutrons are absorbed the chain reaction stops. If scientists do not limit the number of neutrons born after fission, then acceleration will occur (i.e., a nuclear explosion will follow).

The First Nuclear Reactor

The creators of the world's first nuclear reactors faced a difficult task. The fissionable uranium-235 content of natural uranium is negligible at only 0.71%. Most natural uranium is uranium-238 which has a very low probability of fission. The probability of connecting a neutron with the nucleus leading to fission increases significantly when the high temperature of fast-fission neutrons brought about by decelerating neutrons is decreased to almost room temperature.

Therefore, one of the first questions to arise during the development of the nuclear reactor was which direction to follow to find an element moderator: for example, whether to use light (conventional) water, heavy water with deuterium instead of hydrogen to avoid excessive absorption of neutrons in hydrogen, or maybe graphite, beryllium, or paraffin?

Physicists found there were two substances that could slow down neutrons: heavy water (deuterium) and graphite. Heavy water (deuterium) was given priority. Such substances can slow down neutrons to increase the probability of their interaction with uranium, but that would not absorb neutrons. In Nazi Germany heavy water was the focus. A fertilizer plant in Norway seized by the Nazis was producing sufficient heavy water. However, having learned about it the Allies took all the heavy water, transported it to England by sea, and destroyed the plant. Nazi Germany in 1940 was very close to getting division neutrons to react only to realize that it was missing the vital ingredient: heavy water. The design of their first reactor was fascinating. The fuel elements looked like chains of "sausages" made of uranium. By suspending them inside the reactor on a grid a strange effect occurred: neutrons were more selectively absorbed by uranium and reached the fissile nuclei more effectively. If something went wrong the sausage could be released and fall to the floor creating a homogeneous reactor zone in which neutrons were used ineffectively and the reactor would stop. It was at this point that Nazi Germany realized it did not have enough heavy water for critical assembly of the nuclear reactor.

In the United States scientists chose to investigate graphite and heavy water, but gave graphite priority. In December 1942 a team led by the Italian scientist and Nobel Prize winner in physics Enrico Fermi launched the world's first uranium–graphite nuclear reactor called Chicago Pile 1 (CP-1). This reactor was then dismantled and moved to the Argonne National Laboratory (ANL) in Illinois. In 1944 the world's first heavy-water nuclear reactor, the SR-3, was launched at the ANL. A crucial step was the launch of the world's first fast-neutron reactor (without retarder) called Clementine (CP-4) in which liquid mercury was used for the first time as a coolant at the ANL.

Creation of the Atomic Bomb

It was a short step from building a nuclear reactor to building an atomic bomb. However, high-concentration fissile material was needed to create an atomic bomb. The percentage content of fissile material of uranium-235 needed to be significantly increased compared with natural uranium (in other words, uranium-235 needed to be enriched to 93%). Enrichment was found to be extremely energy-intensive and involved a filter made of thousands of membranes with microholes 1 Å in size through which gaseous molecules of UF_6 could pass.

How does such a filter work? $U^{235}F_6$ is only 1% lighter than $U^{238}F_6$. Nevertheless, if this gas made up of two isotopic components that differ in atomic weight by only 1% is passed through thousands of membranes a thousand times, then enrichment can be raised first to 4% and then up to 20%. That is about half the energy costs needed to achieve the 90% required to build a bomb. In the Soviet Union, for example, when the enrichment plant was turned on such a "diffusion" of highly enriched uranium resulted in a power outage in the entire European part of the country.

Another way is centrifuge enrichment. If a centrifuge is rotated at a rate of thousands of rotations per second, then the centrifugal forces of $U^{235}F_6$ and $U^{238}F_6$ can be separated by weight to reach the required enrichment of uranium-235.

The third way to produce fissile material was first implemented in the United States after 1943 at the Hanford Site. At this reprocessing plant plutonium was separated from irradiated uranium from already created nuclear reactors. Although uranium-238 has a very low probability of nuclear fission in a reactor, when absorbed by a neutron it turns into short-lived uranium-239, which after two beta decays turns into the artificial (naturally absent) fissile element of plutonium-239. In terms of its nuclear properties plutonium-239 is a more attractive material than uranium-235 when it comes to creating an atomic bomb.

On July 16, 1945 the first bomb codenamed "Trinity" was detonated in the United States. This explosion is considered the beginning of the nuclear era. Under the leadership of Oppenheimer nuclear weapons were also developed using uranium-233, a second artificial fissile isotope. Uranium-233 is formed in the process of irradiation by neutrons of thorium, a "raw" isotope existing in nature in the form of a single thorium-232 isotope. In other words, thorium does not contain fissile isotopes in nature and can be initially used in a nuclear reactor only with a "fuse" in the form of uranium or plutonium with subsequent production of uranium-233.

The testing of nuclear weapons concentrated the efforts of scientists, engineers, and technologists in Allied countries both during and after the war. The priority was survival that could only be achieved by creating both arsenals of nuclear weapons and their means of delivery. However, the development of nuclear energy for peaceful purposes and the advent of a golden age of humanity with energy prosperity was sidelined by the development of nuclear weapons in the United States and the Soviet Union.

Research School on Nuclear Physics in Russia

Nuclear physics developed in Russia in 1900 as a result of research into the use of radioactive minerals by Russian scientist Vladimir Vernadsky

(1863–1945). Such research involved Vladimir Vernadsky taking part in multiple expeditions to the Ural, Caucasus, and Central Asia regions where radium had been found in ores. After the Great October Socialist Revolution in 1917 the reserves of ore containing radium that remained in St. Petersburg were transferred to the newly created department of the Academy of Sciences. The reserves were eventually evacuated to Vyatka province where the Russian chemist Vitaly Khlopin (1890–1950) came up with an original method to extract radium from the ore. Later, in 1927 the Russian scientist Semenov came up with the theory of chain (branched) reactions. Radium contained in the ores was extracted with the help of the Ferghana Company in 1926–1927. Khlopin was deputy to Vernadsky, later became director of the Radium Institute in the Soviet Union, and then was promoted Head of the Uranium Commission created in 1940.

Famous scientist Georgy Gamov (1904–1968), a student of Khlopin, theoretically explained alpha decay by way of quantum tunneling. Gamov was one of the earliest advocates of the Big Bang theory, which he went to develop. He was also known for his work on molecular biology. Based on Georgy Gamov's theory of alpha decay two researchers at the Cavendish Laboratory, John Cockcroft (1897–1967) and Ernest Walton (1903–1995), in 1932 split the core of lithium into two alpha particles demonstrating that further nuclear transformations could be brought about by bombarding atoms with accelerated protons. Later they both received the Nobel Prize in physics in 1951 for their theory of "atom smashing."

It was also in 1932 that James Chadwick (1891–1974), another student of Rutherford, described the structure of the nucleus and the neutron. He was awarded the 1935 Nobel Prize in physics for his theory. Chadwick planned to continue in the lab of Hans Geiger in Germany after training with Rutherford, but as a result of the outbreak of the First World War he ended up as a prisoner of war kept in the stables of the Potsdam hippodrome. At that time the Berlin company Auergesellschaft had stocks of thorium used in the toothpaste Doramad as a bleach from the 1920s through to the Second World War. James Chadwick negotiated with his guards to get hold of the radioactive dental bleach to continue conduction of his experiments. He spent four years in the Ruhleben internment camp in Germany during the First World War. During the Second World

War, after his Manchester lab was destroyed by the Luftwaffe, Chadwick continued his research to build an atomic bomb as part of the Tube Alloys Project. After merging the British and Canadian Tube Alloys Project with the American Manhattan Project under the Quebec Agreement, he became a chief of the British Mission to work at the Los Alamos Lab and in Washington, DC. For his achievements in physics, particularly for his enormous contribution to building an atomic bomb, he was knighted in Britain in 1945.

Under the Bolsheviks 10 leading research institutes in physics were established in major Russian cities (particularly in St. Petersburg) to boost scientific research into atomic energy. To get an in-depth understanding of the discoveries made by research into nuclear energy and to raise the level of expertise quickly, many prominent Russian physicists, such as Abram Ioffe, Pyotr Kapitsa, and Kirill Sinelnikov, were encouraged by the new Bolshevik regime to study and to carry out research abroad (particularly, in the famous European labs led by Rutherford and Chadwick, Bohr, and Röntgen).

However, after decades of conducting joint research in Europe the Soviet scientists returned home to Russia. For example, the famous scientist Abram Ioffe rejected the proposal of Röntgen to continue working in Germany and came back to Soviet Russia to create research institutes in physics in a number of large industrial cities in Russia.

Indeed, the Soviet Union's Research School of Nuclear Physics developed from the school set up by Abram Ioffe, director of the Institute of Physics and Technology in St. Petersburg. He brought together young and talented students, organized internships with Rutherford. In fact, many future leaders of the Soviet nuclear project, such as Khariton and Landau, went to this school. Thus, until the beginning of the Second World War the development of nuclear physics in the Soviet Union took place in parallel with research in the West (particularly, in Europe). However, the establishment of fascist regimes in Italy, Germany, and Spain was the catalyst for scientists from European countries (especially, from Germany) to emigrate.

Before the Second World War research in nuclear physics in Soviet Russia was concentrated in the central institutes of the Academy of Sciences: in Leningrad at the Institute of Physics and Technology and at the First

Radium Institute; in Kharkov and Moscow at the Institute of Physics and Technology. As early as 1940 a number of the most prominent Russian scientists, such as I. V. Kurchatov, L. I. Rusinov, G. N. Flerov, and Y. B. Khariton, sent the Presidium of the Soviet Union and the Academy of Sciences a letter titled *On the use of uranium energy in chain reactions.* In September 1940 the Uranium Commission was established, which led to approval being given to the first Soviet uranium project.

In 1943, during the Second World War, the Soviet State Defense Committee under the leadership of Chairman of the Soviet Union Joseph Stalin adopted a resolution to organize work on the use of atomic energy for military purposes to protect the country from an external enemy. Supervision of the nuclear problem was entrusted to Molotov, and Lavrenty Pavlovich Beria was tasked with providing military and scientific intelligence. Igor Kurchatov was charged with scientific management of the problem. On August 20, 1945 the Soviet State Committee for Defense established the Special Committee, a new body to manage work on the atomic project. It was entrusted with organizing all activities relating to the use of atomic energy in the Soviet Union: research work, exploration of deposits and uranium mining in the Soviet Union and beyond, creation of the nuclear industry and nuclear power plants, as well as development and production of atomic bombs. The last task was key to implementing the Soviet nuclear project.

What the Special Committee achieved under the leadership of Beria was impressive: in December 1946, just a year after the end of the Second World War, Europe's first nuclear reactor called the F-1 was launched. The F-1 still works at the Kurchatov Institute. All the institutions and participants involved in the Special Committee were daily aware of Beria's exceptional style of leadership and the way in which he organized work likening it to the style of leadership and operational management of the defense industry during that part of the Second World War called the Great Patriotic War (1941–1945) but in a peaceful environment. It was necessary to involve leading specialists from various fields of science and technology, such as metallurgists, mechanics, chemists, biologists, textile workers, and glass specialists, to provide solutions to the multiple problems regarding development of a nuclear reactor to produce a nuclear bomb. This was the time the project Operation Unthinkable proposed by British Prime Minister Churchill became known to the Soviet people. The task of developing a nuclear reactor was complex and could only be solved by uniting as many people as possible from the fields of science and technology.

In-depth research into uranium began in 1942 during the Second World War. Since the United States was not as involved as Russia in the war it was able to become technologically more advanced than the Soviet Union. At the Potsdam Conference, which began on July 17, 1945, the 33rd US President Harry Truman told Soviet leader Stalin that the United States had developed weapons of unspeakable power. He backed this up by pointing out that the day before the conference the Americans had tested a plutonium bomb called "Little Thing" in New Mexico.

Stalin was aware and ready to respond to the threat posed by Churchill and Truman to destroy the Soviet Union. The threat involved a preemptive nuclear attack on at least 20 major cities and industrial locations. The only factor preventing nuclear attack was the fact that the United States/United Kingdom did not have enough materials for such a nuclear attack on the Soviet Union. By 1946, although the Soviet Union was not the first country to create or test a nuclear bomb, it was the first country to develop the bomb to protect itself from being destroyed.

Thus, the atomic project from the perspective of the Soviet Union involved creating a nuclear shield to protect its people from mass destruction. Another prominent feature of the nuclear project in Soviet Russia was its peaceful application of nuclear energy. The Soviets were responsible for the first nuclear-powered icebreaker, the first nuclear power plant created solely for peaceful purposes, and the first medical application of the use of nuclear energy. This completely accorded with the ideas proposed by the founder of nuclear physics, the great Russian scientist Vladimir Vernadsky who wrote in his book *Biosfera* (1926): "Living matter gives the biosphere an extraordinary character, unique in the universe… Cosmic energy determines the pressure of life that can be regarded as the transmission of solar energy to the Earth's surface… Activated by radiation, the matter of the biosphere collects and redistributes solar energy, and converts it ultimately into free energy capable of doing work on Earth… A new character is imparted to the planet by this powerful cosmic force. The radiations that pour upon the Earth cause the biosphere to take on properties unknown to lifeless planetary surfaces, and thus transform the face of the Earth… In its life, its death, and its decomposition an organism circulates its atoms through the biosphere over and over again" (http://vernadsky.lib.ru/).

Igor Kurchatov was appointed Head of the project and would be assisted by Lavrenty Pavlovich Beria and Yulius Khariton. As early as June 1946 Beria was able to provide Kurchatov with secret material describing in detail the blueprints of the American atomic bomb dropped on Hiroshima and Nagasaki in Japan. In 1947 the Council of Ministers of the Soviet Union determined the main objectives of the program in testing the first Soviet nuclear bomb, which carried the symbol РДС-1. In addition, the decision was made to build Plant No. 816 in the Tomsk region in Siberia. This plant would become the largest plant in the Soviet Union for the production of weapons-grade plutonium and highly enriched uranium. Today it is the Siberian Chemical Plant in the city of Seversk in Tomsk. In July 1949 the Semipalatinsk test site was constructed and equipped in readiness to detonate the first Soviet plutonium bomb. On August 29, 1949 the first Soviet nuclear bomb, the РДС-1, with a capacity of 20 kilotons of TNT (the trotyl equivalent) was detonated in Semipalatinsk. The balance of power was restored on the planet. More than 3500 people were awarded government awards for their contribution to this victory.

The first Soviet experimental nuclear reactor constructed at Laboratory No. 2 of the Academy of Sciences of the Soviet Union was successfully launched on December 25, 1946. On August 29, 1949 a test of the first Soviet atomic bomb, the РДС-1, was conducted at a specially constructed test site in the Semipalatinsk region of Kazakhstan. The successful test on September 24, 1951 led to the production of atomic bombs of this type being mastered by the end of that year. On January 1, 1952 the Soviet Union had 35 atomic bombs, 29 of which were РДС-1 and 6 were РДС-2. On August 12, 1953 a thermonuclear weapon was exploded in the Soviet Union. Thus, the initial task of creating nuclear weapons in the Soviet Union and eliminating the nuclear monopoly of the West (particularly, of the United States) was solved. Later, during the Cold War more than 1500 nuclear tests were conducted and more than 85,000 nuclear weapons were created in the United States and the Soviet Union as a result of the global nuclear arms race. An enormous amount of money was spent on the construction of more than 500 nuclear submarines that had about 1000 nuclear reactors and missiles with nuclear charges onboard.

Nuclear Project in Soviet Russia: Creation of a Nuclear Shield

Despite the devastation of the ongoing war Stalin had already ordered work to start on the development of nuclear weapons in 1942. The first thing to do was select a leader for the Russian nuclear project who could combine the talents of an eminent scientist and an eminent organizer. It seemed logical to assign a high-ranking scientist or an academician such as Ioffe, Khlopin, Kapitsa, or Semenov. However, the 39-year-old doctor of physical and mathematical sciences Igor Vasilievich Kurchatov was selected to lead the uranium project of the Soviet Union on the recommendation of Director of the Leningrad Physico-Technical Institute Academician Abram Ioffe.

Igor Kurchatov—Father of the Soviet Nuclear Project

The scientific school of Abram Fedorovich Ioffe is unique in Soviet science. Under his leadership the nation produced a galaxy of famous physicists such as P. L. Kapitsa, I. V. Kurchatov, A. P. Aleksandrov, N. N. Semenov, Y. B. Khariton, Y. B. Zeldovich, L. A. Artsimovich, and I. K. Kikoin. Why was Kurchatov selected as the leader of the nuclear project? Ioffe knew him as an outstanding leader and purposeful scientist who was able to organize and bring work started from scratch to completion. Igor Kurchatov was aware of the importance and grandeur of the task assigned to him and the measure of responsibility that he assumed. By consenting to lead the work on creating an atomic bomb in the most challenging period of the Second World War he demonstrated the courage of a great man and a great scientist.

Kurchatov was not naive. He understood perfectly well what he was undertaking and he knew for sure that from that moment on he would be subject to the control and pressure of state security professionals in view of the extreme secrecy and urgency of the project. No other Russian scientist of the twentieth century had such a massive task, which involved creating an entirely new branch of science called nuclear power in such an extremely short time and under extremely harsh external conditions. As Head of the uranium project Kurchatov had to develop a strategy and the tactics needed to involve the country's leading scientists in the development of entirely new technologies that could be used to obtain fissile materials, to organize the work of multiple R&D organizations, to arrange the construction of uranium mines and industrial enterprises, and to complete the task of creating a nuclear shield around Russia. In other words, to organize from scratch the whole cycle of scientific, financial, and industrial work.

In 1946 in Laboratory No. 2-КБ-11, the most secret branch working on Kurchatov's nuclear project (located in woods in the small town of Sarov), an atomic bomb was being developed under the leadership of talented scientist Yulius Khariton.

A few years earlier, in December 1942 under the supervision of American scientist Robert Oppenheimer, Head of the Los Alamos laboratory, in collaboration with Italian scientist Enrico Fermi the world's first nuclear boiler was launched in the United States. In September 1944 in partnership with Hanford a new reactor capable of processing plutonium was launched. Immediately after collecting enough plutonium the world's first atomic bomb codenamed "Trinity" was constructed and successfully tested in the Alamogordo desert in July 1945. The world's second atomic bomb "Little Boy" and third atomic bomb "Fat Man," both made in the United States, were dropped on civilians of the two Japanese cities of Hiroshima and Nagasaki in August 1945.

The long-awaited end of the war with Germany allowed the Soviet Union to intensify work on the uranium project. The atomic bombing carried out by America of Japanese cities in August 1945 only served to sharpen the Soviet Union's focus on this project. What is more, on March 5, 1946, six months after the first atomic bombings, the famous Fulton speech of Winston Churchill marked the beginning of a new "war" declared against the Soviet Union: the Cold War. According to the plans of Washington and its allies the Cold War could not turn into a real war any earlier than the end of 1949 since even the United States with its atomic monopoly could not shorten the period of preparation for the war.

There was no need to hurry though because it was believed there was no way the Soviet Union would be in possession of atomic weapons before the mid-1950s.

In response to these threats a body called the Emergency Control Authorities for the Atomic Project was set up in Russia. Lavrenty Pavlovich Beria, Head of State Security, was appointed as Head of the Russian Nuclear Project in collaboration with Kurchatov acting as the Scientific Director of the project. Work on the Russian nuclear project was intensified.

In December 1946 the first Soviet atomic boiler called the F-1 was launched at Laboratory No. 2. In the summer of 1948 the first industrial atomic reactor processing plutonium called the A-1 ("Anushka") began operations in the secret city Chelyabinsk-40. By August 1949 a full-fledged explosion of the first Soviet atomic bomb took place at the Semipalatinsk test site.

Thus, in the first two years after the war the Soviet Union, despite suffering terribly in the war and having lost much of its industrial potential, was able to create a colossal nuclear industry from scratch. The creation of a nuclear industry, a nuclear shield for the Soviet Union, and establishing nuclear parity in the world was an incredible feat of Soviet scientists and political leaders.

In 2017 State Duma Deputy MP Nikolai Zemtsov presented a report revealing true statistics about the losses of the Soviet Union in the Second World War citing data from the archives of the USSR State Planning Committee. According to the data:

- The economic damage wrought on the Soviet Union by the war was astronomical. Total decline in the population of the Soviet Union during the Great Patriotic War (1941–1945) was 52,812,000. Of these, irretrievable losses as a direct result of the war were 19,000,000 military personnel and about 23,000,000 civilians. Total natural mortality of soldiers and civilians over this period could have been up to 10,833,000 (including 5,760,000 children who died under the age of four). Therefore, irretrievable losses of the population of the Soviet Union as a result of the war amounted to 42,000,000 people. On November 2, 1942, by decree of the Presidium of the Supreme Soviet of the USSR, an

Extraordinary State Commission was set up to establish and investigate the atrocities perpetrated by German fascist aggressors and their allies as well as damage caused to citizens, collective farms, public organizations, state enterprises, and institutions of the Soviet Union during that part of the Second World War called the Great Patriotic War. Following the war this commission published the following figures. Nazi German invaders and their allies destroyed 1710 cities, 70,000 villages, and deprived about 25,000,000 people of shelter. They destroyed about 32,000 industrial enterprises, 84,000 schools and other educational institutions, and plundered 98,000 collective farms. They destroyed 4100 railway stations, 36,000 communications enterprises, 6000 hospitals, 33,000 polyclinics and dispensaries, 82,000 primary and secondary schools, 1520 secondary specialized educational institutions, 334 universities, 43,000 libraries, 427 museums, and 167 theaters. Losses to agriculture included 7,000,000 horses; 17,000,000 head of cattle; and tens of millions of pigs, sheep, goats, and poultry (either looted or destroyed). Damage to the transportation system of Russia included the destruction of 65,000 kilometers of railway track, 13,000 railway bridges, 15,800 locomotives, and 428,000 wagons. About 1400 ships involved in maritime transport were damaged or stolen. German companies, such as Friedrich Krupp and Co., German Hering, Siemens Shukert, and IT Farbenindustry, committed robbery in the occupied territory of the Soviet Union.

- The material damage caused by German fascist invaders of the Soviet Union amounted to about 30% of its national wealth and in areas under occupation about 67%. Of the total direct economic losses incurred by five of the countries (Soviet Union, United States, United Kingdom, France, and Germany) involved in the war the Soviet Union accounted for 57%. It is noteworthy that the direct economic losses of the Soviet Union were 2.7 times greater than those of the aggressor Germany. This is not surprising since the Third Reich pursued a carpet-bombing strategy in Russia, something Soviet Russia never did. The report of the Extraordinary State Commission was presented at the Nuremberg Trials in 1946.

Nevertheless, the new nuclear power industry emerged in Soviet Russia just a few years after the end of the Second World War. The efforts of hundreds of thousands of people, the billions of rubles spent, and years of hard inhuman work without days off (let alone vacations) all in total secrecy and under constant vigilance were the legacy. For example, Kurchatov could not even step outside his car (let alone the Institute) for years without being followed 24/7 by a bodyguard or a state security officer. Tests of the first atomic bomb were successful. From that day on, the Soviet Union had a nuclear shield.

Moreover, by 1949 scientific nuclear centers such as Chelyabinsk-40, Arzamas-16, and Obninsk were merged under the leadership of Stalin, Beria, Kurchatov, and Khariton and would in time become large leading research institutes.

Almost seven decades ago, on October 29, 1949 the Presidium of the Supreme Soviet of the USSR issued four top-secret decrees awarding 845 people the highest honors: Hero of Socialist Labor, Order of Lenin, and the Red Banner of Labor. All 845 people were directly connected with creating nuclear power in the Soviet Union for the first time.

Ideological opposition and psychological pressure reached such a level during the Cold War that people perceived the possibility of the Soviet Union and the United States using nuclear weapons against each other as an almost inevitable reality. Moreover, nuclear scientists began to create even more destructive and formidable thermonuclear weapons. The "instigator" of this race was the eminent American physicist Edward Teller who still lives in the United States. Kurchatov had to deal not only with setting up the uranium project, but also with the development of the hydrogen bomb (later called a thermonuclear weapon). Since 1948 this project has involved the work of talented scientists such as I. E. Tamm, V. L. Ginzburg, A. D. Sakharov, Y. B. Zeldovich, and Y. B. Khariton. However, the Head of any new nuclear project was always I. V. Kurchatov. One of the most striking traits of I. V. Kurchatov was his ability to inspire people to do a usual, ordinary job regardless of their social rank or level of expertise. He was democratic and had little difficulty speaking the common language of laboratory assistants, prominent academicians, and political leaders. Everybody respected, loved, and followed him. These traits along with his great erudition were fundamental to his phenomenal success and achievements.

After successful testing of the several hydrogen bomb, improving nuclear weapons, and enhancing their production, Kurchatov was finally able to turn his attention to using atomic energy for peaceful purposes in the national economy. Moreover, some of his associates such as A. P. Alexandrov, S. M. Feinberg, and V. I. Merkin had already been working on this. They had busied themselves developing projects for industrial nuclear power plants the first of which was experimental and already operational at the Obninsk Physics and Energy Institute. They were also involved in projects to develop icebreakers and submarines with onboard nuclear power plants. However, Kurchatov's prime objective was creation of a "peaceful atom," a thermonuclear power reactor producing electricity to boost the state economy. Thermonuclear synthesis of light elements had already been implemented in a bomb and now the aim was to be able to control the synthesis to produce electricity. Work on the new project remained extremely secret because it was widely believed in the West that the sole purpose of thermonuclear reactors was as a tool for plutonium collection.

In 1956 Kurchatov was among the entourage accompanying Chairman of the USSR Council of Ministers N. S. Khrushchev on a visit to the United Kingdom, where he delivered two reports. One was about the development of atomic energy in the Soviet Union and the other caused a sensation dealing as it did with thermonuclear fusion research. Kurchatov insisted before the trip that secret peaceful work on thermonuclear reactors should be shared with the West and that all parties in the Cold War should take steps toward rapprochement and collaboration in global science.

Work on the uranium project under the leadership of Kurchatov had led in 17 years to nuclear and thermonuclear weapons, the world's first nuclear power plant, nuclear-powered icebreakers and submarines, and to national nuclear science becoming a future world leader.

The twenty-first century will have to face up to the danger of nuclear weapons proliferating as the number of states possessing nuclear weapons, the so-called nuclear powers, is constantly growing. Aside from the United

States and Russia a number of countries possess nuclear weapons: United Kingdom, France, China, India, Pakistan, Israel, and North Korea (and maybe Iran).

Yulius Khariton—Scientific Director of the Soviet Nuclear Bomb

In 1926 Julius Khariton and Zinaida Walta carried out a series of experiments that Khariton very sparingly calls a study of the oxidation of phosphorus vapor, although it had far-reaching consequences. At that time N. N. Semyonov was engaged in detailed research of these processes and was justified in doing so by discovering branched chemical chain reactions in 1927–1928 for which he was awarded the Nobel Prize.

At the turn of the twentieth century the Russian Empire could not be ranked among the leading countries in science. This meant the scientific physical school had to be built from scratch in Soviet Russia. However, revolutionary changes in physics throughout the world at the beginning of the twentieth century created a situation in which the new generation of physicists could more quickly and more naturally perceive new physical theories than physicists whose outlook was formed earlier. The scientific revolution equalized the chances of all those who had just entered science at that time. All that was needed was talent and access to scientific centers. Political events in Russia at the beginning of the twentieth century had far-reaching social consequences since the powerless strata of the population suddenly had the possibility of social mobility as a result of higher education becoming available to all. In the first decade of its existence the Soviet Union faced a shortage of qualified personnel and made significant efforts to attract young people of working and peasant origin to higher education institutions and to train up politically reliable personnel.

Between 1939 and 1941 Julius Khariton and Yakov Zeldovich not only made the first theoretical calculations of the uranium chain reaction (which turned out later to be very inaccurate), but also demonstrated the possibility of such a reaction.

During the war Khariton's value as a unique specialist in the field of explosives grew ever more important. In 1943 the famous Laboratory No.

2 of the USSR Academy of Sciences was established under the leadership of I. V. Kurchatov. Fully aware of Khariton's unique abilities Kurchatov suggested he take part in the Soviet nuclear project. Thus, Julius Khariton got into the nuclear project from the very first days of its creation, and in 1946 was appointed Scientific Director of the project. He remained in this position until 1992. Julius Khariton was unique in combining the specialties of experimental physicist and theoretical physicist and being able to control all stages of creation of the "product"—from the theoretical idea to its practical implementation.

Khariton's further scientific and organizational activities took place behind the veil of secrecy. In 1949 an atomic bomb was successfully tested in the Soviet Union, and in 1953 and 1955 the first sample hydrogen bombs were tested as well. The state appreciated the merits of Khariton very highly and in 1953 he was elected to the Academy of Sciences of the Soviet Union. During this period of his life Khariton was Head of a scientific project and Leader of a large group of prominent scientists.

In secret documents relating to creation of the USSR's first nuclear bomb the term "product" was used. Officially called the "special jet engine" (in Russian "реактивный двигатель специальный") it was abbreviated as "РДС." Thanks to Julius Khariton the "РДС" abbreviation was promptly unofficially decrypted as "Россия делает сама" or "Made in Russia." Although there was a fair degree of irony in this because everyone knew how much intelligence Russian scientists had obtained from foreign science, there was also a large degree of truth. After all, could it not be coincidental that the design of the first Soviet nuclear bomb was very similar to the American one simply because the best decisions were taken? Since the laws of physics and mathematics do not have national boundaries it could be argued that the ballistic body and the electronics filling the first bomb were a purely Russian development.

The first Soviet nuclear bomb was detonated at 7 a.m. on August 29, 1949. Almost a month passed before Western authorities recovered from the shock when intelligence about the successful testing of an atomic bomb in Russia reached them. On September 23 US President Harry Truman, who had so recently boastfully informed Stalin about America's successes in creating atomic weapons, issued a formal statement that the same type of weapon was now also available in the Soviet Union.

Strangely enough there was no hurry in Moscow to confirm the statements of the Americans. On the contrary, official representatives refuted the American statement arguing that a device was being worked on in the Soviet Union involving the latest technologies. However, there was official confirmation on November 6, 1947 by USSR Foreign Minister Vyacheslav Molotov who declared that when it comes to atomic bombs no secret survives for long.

Indeed, up to 1947 all information about atomic weapons had been kept a secret by the Soviets, but by the end of summer 1949 it was clear to everyone that the Soviet Union had restored strategic parity with the United States, its main rival. It was only on March 8, 1950 that the Soviet Union officially announced the existence of the so-called "product 501"—the atomic bomb. This announcement was made when the Korean War of 1950–1953 broke out on the Korean Peninsula (June 1950). This resulted in the sovereign state of Korea being divided into two bitter enemies: South Korea and North Korea. Then most people in the world feared this bloody massacre, which claimed the lives of 4 million peaceful Koreans, could develop into a world atomic war based on statements made by US General Douglas MacArthur who was defending South Korea. However, the nuclear war did not happen. Key to this was the creation of nuclear parity—equilibrium in the capabilities of the two nuclear superpowers, Russia and America, to conduct military operations using the atomic arsenal. The formal date when such parity became a reality was Wednesday, March 8, 1950. It was then that Deputy Chairman of the USSR Council of Ministers Marshal of the Soviet Union Clement Voroshilov finally informed the whole world that the Soviet Union possessed its own atomic weapons.

On August 12, 1953 Soviet specialists conducted the first test of a hydrogen bomb, lagging behind the American tests by 11 months (having narrowed the gap that used to be 5 years). The October 30, 1961 detonation in Novaya Zemlya of the Soviet so-called "Tsar Bomba," a thermonuclear "product" with a capacity of almost 60 megatons, signaled the end of Western threats to undertake pre-emptive nuclear strikes on the Soviet Union and/or to declare nuclear war against other countries. Indeed, in just a decade Russian scientists, engineers, and military turned their country from a war-weary power, threatened with destruction by new nuclear weapons, into a powerful nuclear superpower.

Thus, it should be remembered that nuclear parity preserved throughout the world for more than seven decades was brought about by the Soviet Union. Nuclear parity came about as a result of Soviet nuclear scientists in post-war Russia creating a nuclear shield for the Soviet Union and maintaining peace on planet Earth.

Creation of the the "Tsar Bomba," the First Hydrogen Bomb

Research in nuclear physics was first conducted in the Soviet state in the pre-war years. In 1921 the Radium Laboratory (later the Radium Institute) at the Academy of Sciences was set up under the stewardship of Khlopin. In 1935 a beam of accelerated protons was obtained for the very first time at the Radium Institute from the first cyclotron in Europe. In 1939 Zeldovich, Khariton, and Leipunsky substantiated the possibility of a nuclear chain fission reaction in uranium. In September 1940 the USSR Academy of Sciences approved a program of work on the study of uranium fission reactions.

In the midst of the Second World War, on September 28, 1942 the USSR Academy of Sciences was tasked with looking into the feasibility of using atomic energy by splitting uranium nuclei and submitting a report on the possibility of creating a uranium bomb or uranium fuel by April 1, 1943. By August 20, 1945 a special body was set up to manage all work concerning uranium. This was the Special Committee under the State Defense Committee of the Soviet Union consisting of top government figures and physicists. In 1944 the first few kilograms of pure uranium ever found in Eurasia were obtained. In 1946, for the first time on the continent of Eurasia, a self-sustaining chain reaction of uranium fission was carried out under the leadership of Kurchatov in the first ever nuclear reactor.

Two years later such work led to the launch of the first industrial reactor "A" for the production of plutonium at Plant No. 817 (Mayak in Ozersk, Chelyabinsk region). Moreover, on August 29, 1949 the first Soviet nuclear charge (РДС-1) was successfully tested at the Semipalatinsk test site. In 1951 a second atomic bomb was tested, and in 1953 the first Russian thermonuclear bomb (РДС-6s) was tested. Four years later the first nuclear submarine was built (Project K-3).

As early as April 1949 the first heavy-water research reactor in the Soviet Union (or Europe for that matter) was launched leading to a number of significant discoveries being made. In May 1950 the world's first nuclear power plant near Obninsk was established. The station was equipped with a water-cooled uranium–graphite channel reactor called "Atomic Peace" that had a capacity of only 5 MW. The design of the station core was proposed by Kurchatov together with Professor S. M. Feinberg. The chief designer was Academician N. A. Dollezhal.

In 1955 the world's first fast-neutron reactor the BR-1 with zero power was put into operation, followed a year later by the BR-2 with thermal power of 100 KW. The decision to build the first nuclear-powered icebreaker was made on November 20, 1953.

On 30 October 1961 the Soviets detonated a hydrogen bomb (codenamed "РДС-220," "Ivan," or "Vanya"), known by Western nations as the "Tsar Bomba" or "Kuzka's Mother." This relates to First Secretary Nikita Khrushchev's promise to "show the United States a Kuzka's mother" (an idiomatic expression meaning "We'll show you") at a 1960 session of the UN General Assembly. The "Tsar Bomba" was the most powerful nuclear weapon ever created. Tested as an experimental verification of calculation principles and multistage thermonuclear weapon designs, it also remains the most powerful explosive ever detonated. The bomb was detonated secretly but was detected by international intelligence agencies causing panic in military circles in the West. One of the architects of the "Tsar Bomba" was the Soviet physicist Andrei Sakharov. Ironically, he would become famous for his attempts to rid the world of the very weapons he had helped create.

In 1964 the first VVER-1 reactor with a capacity of 210 MW was launched at the Novovoronezh Nuclear Power Plant. In 1973 the world's first fast-energy neutron reactor BN-350 was launched, followed in 1974 by the launch of the first 1000 MW RBMK reactor at the Leningrad Nuclear Power Plant.

The period between 1957 and 1986 was particularly important as a number of large nuclear power plants were built and work on controlled thermonuclear fusion was developed to a very high degree. In 1967 the largest (at that time) accelerator of protons, the U-70, capable of producing energy of 70 billion eV was launched at the Institute of High Energy Physics. The construction of nuclear power plants in Eastern Europe began. Between 1957 and 1967 the Soviet Union built 25 nuclear installations including 10 nuclear power plants, 7 accelerators, and 8 isotope physics laboratories in Warsaw Pact countries in Eastern Europe, in Asia, and in Africa. However, the accident at the Chernobyl Nuclear Power Plant in 1986 in Ukraine slowed down the development of domestic nuclear energy, and in the 1990s the Russian nuclear industry went through a period of stagnation. In December 2007 the State Atomic Energy Corporation ROSATOM was formed.

"Peaceful Atom": World's First Peaceful Nuclear Reactor

In 1953 Dwight D. Eisenhower came up with the idea of establishing an international authority to control the development and use of nuclear technologies. This led to the establishment of the International Atomic Energy Agency (IAEA). The First Geneva Conference on the Peaceful Uses of Nuclear Energy where peaceful uses of nuclear energy would be discussed was held under the auspices of the United Nations and represented the first practical step to setting up the IAEA. At this conference a report on the world's first nuclear power plant in the Soviet Union launched in 1954 in Obninsk caused a sensation. This was a uranium–graphite reactor cooled with enriched uranium water.

After that the number of nuclear power plants around the world began to grow rapidly. All plants were based on weapons-grade nuclear projects right from the fuel base, all the industrial infrastructure (from uranium mining to nuclear fuel production and use in the reactor), to the training, knowledge, and experience of specialists. Even the new types of nuclear power plants that began operation were little more than converted military developments.

By the mid-1980s up to 40 nuclear units were being built in the world each year. Everything was going fine until 1979 when the most massive accident in the history of commercial nuclear power occurred. This was in the United States at the Three Mile Island Nuclear Power Plant and became a national nuclear disaster.

Then, in 1986 the accident at the Chernobyl Nuclear Power Plant in the Ukraine occurred. This turned into a nuclear disaster with huge sociopolitical consequences. Although the problem mainly concerned the Soviet Union since reactors of this type were only built there, the negative attitude of people in European countries increased dramatically. Seven countries decided to ban the development of nuclear power. Revising nuclear power plant safety and developing a "nuclear safety culture" became fundamental to the continuation of nuclear power.

Two international projects were organized in 2000 to address the nuclear crisis. The Innovative Nuclear Power Project (INPRO) proposed by Russia was developed under the auspices of the IAEA and brought together specialists from developed and developing countries. The purpose of INPRO was to define the development strategy and requirements for future nuclear power.

The other project involved the development of fourth-generation reactors for future nuclear power plants. It was aimed at creating reactors that not only solved problems relating to further development of nuclear power, but also put safety, economic efficiency, unlimited development resources, waste and spent fuel management, and non-proliferation at the forefront. This project proposed by the United States involved only the 10 most developed western countries in the field of nuclear energy.

Unfortunately, the situation has deteriorated dramatically as a result of the largest nuclear accident in the twenty-first century, which took place at the Fukushima Nuclear Power Plant in 2011 in Japan, one of the most industrially advanced countries in the world.

References

Brodov, V. (1984). *Indian Philosophy in Modern Times.* Moscow: Progress Publishers.

Dasgupta, S. (1922). *History of Indian Philosophy.* Cambridge: Cambridge University Press.

Freeman, K. (1948). *Ancilla to the Pre-Socratic Philosophers.* Cambridge, MA: Harvard University Press.

Radhakrishnan, S. (1923). *Indian Philosophy.* New York: Oxford University Press.

7
Global Nuclear Policy Development

The history of nuclear energy covers a period of more than half a century, and during that time, it has already become a traditional branch of energy. Currently, 31 countries operate nuclear power plants. By the beginning of 2017, there are 451 power reactors in the world (not including those shut down for a long time) with a total capacity of 392,521 MW, 60 reactors were under construction. The vast majority of nuclear power plants are located in Europe, North America, Far East Asia, and the territory of the former USSR, while in Africa there are almost none, and in Australia and Oceania, there are none at all.

The share of electricity production at nuclear power plants in some countries reaches high values, for example, in 12 countries it exceeds 30%. On the other hand, in some countries, the share of nuclear energy in the energy balance is insignificant, for example, China is one of the leaders in installed capacity; however, nuclear power plants provide about 3–4% of the country's electricity. The United States is the global leader in installed capacity, but nuclear power makes up only 20% of the total energy balance of this country. The world leader in terms of the share of total output is France (second place in terms of installed capacity), in which nuclear power is a national priority.

D. Basu and V. W. Miroshnik, *The Political Economy of Nuclear Energy*,
https://doi.org/10.1007/978-3-030-27029-2_7

At the same time, there are opposing trends in the world, expressed in stagnation and even abandonment of nuclear energy. As some leaders of the nuclear power industry (USA, France, and Japan), and some other countries, have closed a number of nuclear power plants. Italy became the only country to close all existing nuclear power plants and completely abandon nuclear power. Belgium, Germany, Spain, and Switzerland implement a long-term policy to abandon nuclear energy. Azerbaijan, Georgia, Lithuania, and Kazakhstan abandoned nuclear power, largely due to the collapse of the USSR, with nuclear power plants operating in the framework of the unified Soviet power industry already operating on the territory of the last two countries. Austria, Cuba, Libya, the DPRK (North Korea), and Poland for political, economic, or technical reasons stopped their nuclear programs before launching their first nuclear power plants, although Poland does not exclude the possibility of building nuclear power plants again in the long term. Previously, Armenia had abandoned nuclear power, but then its only nuclear power plant was put back into operation. The Netherlands, Taiwan, and Sweden have planned to abandon nuclear energy, although so far they have suspended such measures. Nuclear power development programs were abandoned in Australia, Ghana, Greece, Denmark, Ireland, Latvia, Liechtenstein, Luxembourg, Malaysia, Malta, New Zealand, Norway, Portugal, and the Philippines. Prospects for the construction of new nuclear power plants in some countries also raise doubts.

US nuclear power is the largest in the world in terms of energy generated. According to data for September 2018, 98 nuclear reactors with a total capacity of 99.3 GW (gigawatt) operate in the United States, which produce 20.05% of the country's electricity (the average capacity factor over 15 years is more than 90%). Almost all operating nuclear power plants were built in the period 1967–1990, new nuclear power plant projects were launched only in 2013. As of March 2018, the construction of 2 new reactors with a total capacity of 2.5 GW continues in the country. The construction of 2 more AP1000 reactors was decided to be abandoned, due to the double increase in the cost of estimates and the reluctance of the parties to incur additional costs.

Over the years of construction, the payback of nuclear power plants in the United States has come into question. Mass production of shale gas lowered its price almost three times. Atomic energy began to lose gas power plants.

Global Nuclear Policy Development

There is no such contradictory attitude to any other sphere of energy as to nuclear generation. The chapter considers the whole range of conditions existing in different countries for the development of nuclear energy—from the most favored nation treatment to the complete ban.

Atomic energy uses dual-use technologies, is associated with increased risks, huge capital, and high scale science. Unlike most other industries in which conditions for investors are good enough if the state minimizes interference and taxes only by monitoring the integrity of competition and compliance with norms, the participation of the authorities in the nuclear sphere is essential. Therefore, the "nuclear" investment climate is determined primarily by the state's attitude and the forms of its influence on the industry. With all the diversity of options, all countries can be divided into several groups, the boundaries between which are blurred in places.

The first group includes countries with a pronounced manifest policy, state protectionism with respect to nuclear energy. The second group includes countries with a more or less neutral attitude toward nuclear energy. In them, the state support of nuclear power is very limited; as a consequence, its development occurs mainly according to the laws of economic expediency, proceeding from the changing market conditions. The third group includes countries in which nuclear power is placed in less favorable conditions than other energy sources. In each group, there are both moderate and radical options.

In the countries of this group, the state determines the progressive development of nuclear power, regardless of market conditions. That is, state protectionism in the industry achieves the highest degree, manifesting itself in different ways. For the nuclear industries of countries with state protectionism, in its maximum variant, which can be called "nuclear absolutism," some standard features are characteristic. First of all, this is the complete domination of the state in all or almost all spheres of the industry. As a rule, the lion's share of sectoral resources is concentrated in the hands of one or two state corporations or companies with predominant state participation, which provide almost a state monopoly: in Russia it is ROSATOM, in India—NPCIL and BHAVINI, in France—EDF and Areva, in South Korea—KEPCO and its subsidiary structure KHNP.

If the non-controlling share of private investors is sometimes allowed in the listed holdings, the main R&D is usually withdrawn into separate organizations (typically non-profit), completely and directly controlled by the state, for example, CEA in France, SRC "Kurchatov Institute" and other nuclear centers in Russia, Bhava Research Centres in India, and KAERI in the Republic of Korea. The role of private structures in this area is insignificant, in contrast to some countries of the second group.

In a number of cases, along with federal, a significant role belongs to regional and municipal authorities. Of the countries under consideration, this primarily concerns China: Guangdong, Fujian, Guangxi, Zhejiang, Hubei, and some cities have significant shares in nuclear projects in this country.

The role of private companies in the industry is limited, albeit to varying degrees. In France, India, and Russia, their share is insignificant, while in China and South Korea, the presence of non-state structures is more noticeable. However, even here, they participate indirectly—through companies with dominant state control. Thus, in South Korea, about 49% of KEPCO holding shares belong to non-state entities, including foreign ones. In turn, KEPCO through its "daughter" KHNP undividedly controls all the country's nuclear power plants.

Atomic Liberalism

In contrast to the group of countries of "nuclear protectionism," atomic liberalism exists in the countries of the market investment climate favorable in the nuclear industry, the state protects it from the negative market conditions and does not set mandatory parameters for the development of nuclear power.

The intervention of state structures in the industry is limited mainly to supervisory and supervisory functions in matters related to nuclear safety: certification of equipment, licensing of projects, and the like. The development of nuclear generation depends primarily on market conditions. In the countries of this group, the state can directly support individual projects, but the private sector still plays a decisive role in the development of nuclear energy.

For example, in the United States, the state is guided by a specific energy development strategy, but it has an indicative, scenario character. The White House or some state cannot require private structures to build at their own expense a certain amount of nuclear-generating capacity or choose one or another technology from among the certified ones. The government influences the development of the industry not by policy decisions, as in the countries of "nuclear absolutism," but by means of economic incentives or disincentives. The exception is the questions, one way or another related to nuclear and radiation safety, physical protection, and so on.

In a number of countries of the "liberal group," for example, in Canada, Finland, the United States, regions and municipalities take part in the development of nuclear energy, along with private companies. However, the authorities of this level do not have a monopoly in the industry. The absence of a state monopoly in nuclear energy is another characteristic feature of the countries under consideration. Indeed, regardless of how many players are in the nuclear generation sector, non-state structures occupy an essential place in it. Nor have the spheres of nuclear fuel supply been monopolized by the state, the design and construction of nuclear power plants, the manufacture of the basic equipment of nuclear power plants, and others.

The most striking example is the United States. Although the state is directly or indirectly present in all spheres of nuclear energy—from the supply of technology to the generation of electricity—in the civilian part of the industry this presence is small compared to private structures. So, out of 61 operating in the US nuclear power plant, only three are in federal ownership, the rest belong to dozens of companies in which regional and municipal structures participate. In the civil nuclear fuel cycle, non-state companies also predominate (they, by the way, also fulfill essential functions in the US nuclear weapons complex). In the market of equipment for nuclear power plants, there is competition between foreign and domestic suppliers that are not controlled by the state: Westinghouse and Toshiba, GE-Hitachi, B & W, Areva, Mitsubishi, and others.

An example of the more significant role of state structures is Canada, where the state has abandoned monopoly in a number of sectors of the industry only in the last decade and a half. A significant portion of the assets of the existing nuclear generation is still controlled by the provinces (today it is Ontario, and New Brunswick, formerly Quebec was one of them). However, a significant role is played by non-state structures, which, for example, control the Bruce NPP in Bruce Power, the most powerful in Canada and throughout North America.

In the nuclear fuel cycle of Canada, which is one of the largest in the world, there is also no monopoly: the most important (but not the only) players in uranium exploration and production are Cameco and Areva, and Cameco and GE-Hitachi compete in the upper levels of nuclear fuel production. Finally, at the beginning of this decade, the state (federal government) refused to monopolize the supply of domestic reactor technologies, and SNC-Lavalin Group, an engineering holding, became the most crucial player in this field.

At the same time, the main equipment market for nuclear power plants is open to foreign suppliers, which occupy an important place in it (in niche manufacturing of steam generators, turbine equipment, etc.) and have the opportunity for further expansion, including the supply of alternative reactor designs to Canadian ones.

Another example is Finland, where the state-controlled Fortum joint-stock company owns a significant portion of some (but not all) nuclear projects, but a considerable part of the assets are held by companies that are not controlled by the government. At the same time, the role of the state in supplying equipment for the nuclear industry is insignificant, as in most other countries that do not have their own key nuclear technologies and production base for the production of the main equipment of nuclear power plants.

Also, competition is developed, and the presence of non-state capital in the nuclear power industry of Spain is significant. However, unlike the United States, Finland, or Canada, there are precedents for preferential control of nuclear power plants by foreign investors (a substantial portion of the assets belongs to the Italian Enel). On a similar liberal path, at least for now, Turkey is moving ahead, allowing competition and full control of foreign investors in its future nuclear industry.

The market conditions for nuclear power differ substantially both in the countries of the group under review and in regions (not necessarily administrative ones) within the same country. The most important characteristics are the degree of liberalization of the electricity market and the working conditions of nuclear power plants on it. In the United States, since the liberalization of some regional electricity markets, all construction or completion projects for new nuclear power units have been carried out in regulated regulatory areas where nuclear power plant owners have the opportunity to enter into long-term contracts for large volumes and to pledge a significant portion of their costs and investments in tariffs approved by state regulatory bodies.

Regional differences and differences in market conditions in Canada's electric power industry are also significant, but their impact on nuclear generation is exactly the opposite. For example, the Quebec market has always remained one of the most "regulated" in Canada. In the region, there is a surplus of generating capacities, and mostly cheap hydropower. Against this background, the cost of generating the only "Gentili" NPP operating in this province increased over the years, and in the event of a significant overhaul, which was inevitable, the station would become unprofitable. And the regional authorities decided to abandon nuclear energy; in 2012 the station was closed.

At the same time, in other regions of Canada (Ontario, New Brunswick), where the liberalization of the electricity market has moved further, nuclear generation has survived, although its profitability has suffered from a change in the competitive situation, the need for costly overhauls (for Canadian-designed reactors they are mandatory in about 25–30 years of operation).

Atomic Absolutism

The second peculiarity of some countries of "atomic absolutism" is that the state controls not only the spheres of atomic energy use and the nuclear fuel cycle but also the production of a large part of the equipment. The highest degree of monopoly of this kind is achieved in France, China, and Russia, where companies with predominant state participation control the nuclear power plants themselves, but also a significant share of the production of the main equipment of the nuclear island, and often the construction of nuclear power plants. In France, for example, almost all the technological chain of production of the primary equipment of the primary circuit is concentrated in several enterprises belonging to the state-controlled structures (now—Areva, then, obviously, EDF).

In Russia, with respect to the part of the production chain (manufacturing forgings, reactor housings, steam generators, etc.), there is a competition between two domestic suppliers, one of which (ROSATOM) is fully controlled by the state, the other (OMZ)—in part. Competition is even more developed in China, but even there, domestic suppliers of the main equipment of nuclear power plants remain under primary state control: first of all, such significant companies as China First Heavy Industries, Shanghai Electric, and Dongfang Electric.

The third distinguishing feature of the countries of "nuclear absolutism" is a significant share of state-controlled cash flows (funds of state budgets and state-controlled companies) in the structure of investments in the nuclear industry. In some countries, for example in France, market sources of financing predominate (own means of state companies and loans from the capital market), in others, along with such sources, state guarantees, preferential credits of state financial institutions, state investments, including capital, companies (for example, in Russia).

Also at the same time, there are significant differences between the countries of "nuclear absolutism" in relation to alternative, foreign suppliers of nuclear technologies: if some have practically closed the domestic market for them (Russia, France, and South Korea has advanced in this direction), others, demonstrate the maximum openness of the market (India, China).

At the same time, such countries seek to maximize the localization of production of equipment and construction of nuclear power plants, striving to gain unlimited rights to introduce foreign nuclear technologies gradually.

This process reached the most advanced stage in South Korea, which received almost unlimited rights and technical capabilities for the introduction of Korean reactors based on licensed foreign technologies on the domestic and international markets (in particular, the flagship model of the South Korean reactor APR-1400 was developed on the basis of the System 80+ American company Combustion Engineering, absorbed Westinghouse ultimately).

Almost as far in adapting foreign technologies, China managed to enter, but with reference to a larger number of reactor concepts than South Korea. India is only at the beginning of this path with regard to the construction of light-water reactors, but at the same time has almost unlimited rights to introduce its heavy-water reactor facilities, most of which retain many of the technological features of their Canadian prototypes.

The protectionist model of the industry does not remain unchanged, but in a number of countries, it is noticeably transformed. An example is France. On the one hand, it reached the peak of "nuclear absolutism": almost complete control of the state overall spheres of the industry, the unprecedented role of the NPP in the national energy balance (about 75% of output), a high level of development of nuclear technologies and so on. On the other hand, in recent years the state has refused to build up nuclear energy within the country further, having decided to stabilize its absolute scale at the current level. However, in practice this does not mean abandoning rigid protectionism, but slightly changing its priorities: France shifts the emphasis of state support to increased foreign expansion and the qualitative rather than quantitative development of its own nuclear power.

An example of an even more significant transformation is Great Britain. Historically, the nuclear energy of this country developed within the framework of "nuclear absolutism," but in a few decades, Britain began to move toward liberalization of the industry gradually. To date, the UK has become a model of a kind of market-based model of protectionism, in many respects reminiscent of liberal. However, it cannot be said that having withdrawn from the "state dictatorship" in the industry, the United

Kingdom won on the way to "nuclear democracy," the country lost its independence with regard to a number of key nuclear civilian technologies. One of the reasons for this was a sharp decline in state support for nuclear R&D at the end of the twentieth century; in this indicator, the UK fell behind most large countries, including some that do not have nuclear generation.

Protecting countries should also include countries in which the nuclear-controlled state controlled by the state and supported by it plays a special role in the existing and prospective energy balance, but in recent years the state has done little to implement this strategy, primarily because of a lack of resources. To such countries, "necessarily limited" protectionism is prevailing, for example in Ukraine and the Czech Republic.

The Czech Republic, in its energy strategy approved in mid-2015, made an unambiguous bet on nuclear power, which in the future should become the predominant component of the generation balance: by 2030, it is expected to increase its share from 30–35% to 50–60% national power generation. This could turn the energy of the Czech Republic into one of the most "nuclear" in the world.

However, there is as yet no practical advance in this direction; in 2014 a tender was canceled for the construction of new blocks of the Temelin NPP, and today the state is considering the possibility of the first expansion of another nuclear power plant, the Dukovany. A stumbling block in the implementation of the plans was the state's unwillingness to compensate for part of the financial risks of construction and the subsequent operation of new facilities—through state guarantees or long-term pricing, similar to that adopted in the UK. This problem has not yet found a final solution, despite the willingness of a number of technology suppliers (the Russian VVER-1200 reactor and the American AP1000) to attract foreign investments covering part of the costs.

Similar uncertainty with the renewal of the nuclear park and sources of its financing is typical for Ukraine, which also gives nuclear power an exclusive place in the current and prospective energy balance, but currently does not have the resources for its integrated development. In general, in countries of "reluctantly limited" protectionism, the state is entirely determined to develop nuclear energy but is not yet able to provide it at the expense of its resources. However, in some other similar countries

(Argentina, Hungary, Romania for example) in recent years, the situation has shifted from words to action: framework agreements have been reached with technology suppliers that envisage the construction of new nuclear facilities. In all these cases, it was decided to attract foreign investors or borrowed funds for partial financing of projects.

The NPPs also fit into the competitive markets of some other states of the group under consideration, where gas generation costs much more than in the United States, and the lack of resources and high costs for reducing greenhouse emissions hinder the development of coal. Under such conditions, nuclear power can remain competitive even against the backdrop of the highest degree of liberalization of the electricity market. The example is Finland, whose nuclear power plants remain highly competitive, integration of the country into one of the most liberal electricity markets in the world (Scandinavian), a high proportion of cheap hydro generation, developed intersystem network connections in the zone of this market, and so on.

Another difference of this group from the countries of "nuclear absolutism" is the active participation of non-state structures in R&D. For example, in the United States, a significant part of R&D for prospective reactor designs, nuclear fuel cycles, decommissioning is carried out by non-state entities, as well as subsequent introduction of developed industrial technologies. "Privatization" of the R&D sphere reaches such a degree that non-state companies create dual-use technologies. Thus, the industrial technology of separation of uranium isotopes by means of a laser in this country is created by private structures, with the participation of foreign investors—Japanese and Canadian.

For nuclear power in the countries of the group under consideration, the taxation regime common to the rest of the energy sector is usually used, adjusted for the specifics of the industry (the need for deductions for decommissioning and management of SNF). However, in a number of cases—for example, Spain, which in recent years, has increased the taxation on energy, including nuclear power.

Spanish companies operating nuclear power plants replenished the fund for decommissioning and handling SNF in proportion to the volume of electricity produced—in the approximate range of 0.006–0.008 euro/kWh. In 2013, two taxes were imposed on the handling and storage of spent nuclear fuel, as well as high-level waste. The first is the tax on the unloaded fuel: 2190 euro/kg t. M. discharged from the reactor SNF. The second is the tax on the planned centralized storage of SNF. It is 70 euro/kg t. M. for SNF and 30,000 euro/m for HLW (high-level waste). Together, both taxes are equivalent to approximately 40–50 million Euros per 1 GW of installed capacity per year. Two reasons explain tax changes in Spain. The first is Madrid's intention to shift in full the costs of decommissioning and handling SNF and HLW from taxpayers to energy companies. The second is the need to cover the tariff imbalance that arose in the country's power industry due to significant subsidization of RES, which was not compensated by the previously established tariffs. The increase in the fiscal burden in Spain has significantly affected not only nuclear energy, and this distinguishes it from the special nuclear taxes established in some countries with a biased attitude toward nuclear energy.

In the countries of this group, the state determines the progressive development of nuclear power, regardless of market conditions. That is, state protectionism in the industry achieves the highest degree, manifesting itself in different ways. For the nuclear industries of countries with state protectionism, in its maximum variant, which can be called "nuclear absolutism," some standard features are characteristic. First of all, this is the complete domination of the state in all or almost all spheres of the industry. In China, the CLP Hong Kong group is the co-owner of the Taiwan plant in Guangdong Province (through its subsidiary structure HKNIC, it owns a 25% stake in the company-owner of the station—Guangdong Nuclear Power Joint Venture Co. Ltd.). Another example in the same country: the French EDF owns ~30% of the Chinese company TNPJVC, which is carrying out the project of construction of the "Taishan" NPP. However, the largest ultimate beneficiary of both named stations is the state-controlled (central government and Guangdong province) CGN holding.

The second peculiarity of some countries of "nuclear absolutism" is that the state controls not only the spheres of atomic energy use and the nuclear fuel cycle but also the production of a large part of the equipment. The highest degree of monopoly of this kind is achieved in France, China, and Russia, where companies with predominant state participation control not only the nuclear power plants, but also a significant share of the production of the main equipment of the nuclear site, and often the construction of nuclear power plants.

The third distinguishing feature of the countries of "nuclear absolutism" is a significant share of state-controlled cash flows (funds of state budgets and state-controlled companies) in the structure of investments in the nuclear industry. In some countries, for example in France, market sources of financing predominate (own means of state companies and loans from the capital market), in others, along with such sources, state guarantees, preferential credits of state financial institutions, state investments, including capital, companies (for example, in Russia).

However, at the same time, there are significant differences between the countries of "nuclear absolutism" in relation to alternative, foreign suppliers of nuclear technologies: if some have practically closed the domestic market for them (Russia, France, and South Korea has advanced in this direction), others, demonstrate the maximum openness of the market (India and China). At the same time, such countries seek to maximize the localization of production of equipment and construction of nuclear power plants, striving to gain unlimited rights to introduce foreign nuclear technologies gradually.

This process reached the most advanced stage in South Korea, which received almost unlimited rights and technical capabilities for the introduction of Korean reactors based on licensed foreign technologies on the domestic and international markets (in particular, the flagship model of the South Korean reactor APR-1400 was developed on the basis of the System 80+ American company Combustion Engineering, absorbed Westinghouse ultimately).

Almost as far in adapting foreign technologies, China managed to enter, but with reference to a larger number of reactor concepts than South Korea. India is only at the beginning of this path with regard to the construction of light-water reactors, but at the same time has almost unlimited rights to introduce its heavy-water reactor facilities, most of which retain many of the technological features of their Canadian prototypes.

The protectionist model of the industry does not remain unchanged, but in a number of countries, it is noticeably transformed. An example is France. On the one hand, it reached the peak of "nuclear absolutism": almost complete control of the state over-all spheres of the industry, the unprecedented role of the NPP in the national energy balance (~75% of output), a high level of development of nuclear technologies and so on. On the other hand, in recent years the state has refused to build up nuclear energy within the country further, having decided to stabilize its absolute scale at the current level. However, in practice this does not mean abandoning rigid protectionism, but slightly changing its priorities: France shifts the emphasis of state support to increased foreign expansion and the qualitative rather than quantitative development of its own nuclear power.

An example of an even more significant transformation is Great Britain. Historically, the nuclear energy of this country developed within the framework of "nuclear absolutism," but in a few decades, Britain began to move towards liberalization of the industry gradually. To date, the UK has become a model of a kind of market-based model of protectionism, in many respects reminiscent of liberal. However, it cannot be said that having withdrawn from the "state dictatorship" in the industry, the United Kingdom won on the way to "nuclear democracy," the country lost its independence with regard to a number of key nuclear civilian technologies. One of the reasons for this was a sharp decline in state support for nuclear R&D at the end of the twentieth century; in this indicator, the UK fell behind most large countries, including some that do not have nuclear generation.

Protecting countries should also include countries in which the nuclear-controlled state controlled by the state and supported by it plays a special role in the existing and prospective energy balance, but in recent years the state has done little to implement this strategy, primarily because of a lack of resources. To such countries, "necessarily limited" protectionism is prevailing, for example in Ukraine and the Czech Republic.

The Czech Republic, in its energy strategy approved in mid-2015, made an unambiguous bet on nuclear power, which in the future should become the predominant component of the generation balance: by 2030, it is expected to increase its share from 30–35% to 50–60% national power generation. This could turn the energy of the Czech Republic into one of the most "nuclear" in the world.

However, there is as yet no practical advance in this direction; in 2014 a tender was canceled for the construction of new blocks of the Temelin

NPP, and today the state is considering the possibility of the first expansion of another nuclear power plant, the Dukovany. A stumbling block in the implementation of the plans was the state's unwillingness to compensate for part of the financial risks of construction and the subsequent operation of new facilities—through state guarantees or long-term pricing, similar to that adopted in the UK. This problem has not yet found a final solution, despite the willingness of a number of technology suppliers (the Russian VVER-1200 reactor and the American AP1000) to attract foreign investments covering part of the costs.

Similar uncertainty with the renewal of the nuclear park and sources of its financing is typical for Ukraine, which also gives nuclear power an exclusive place in the current and prospective energy balance, but currently does not have the resources for its integrated development. In general, in countries of "reluctantly limited" protectionism, the state is entirely determined to develop nuclear energy but is not yet able to provide it at the expense of its resources. However, in some other similar countries (Argentina, Hungary, Romania for example) in recent years, the situation has shifted from words to action: framework agreements have been reached with technology suppliers that envisage the construction of new nuclear facilities. In all these cases, it was decided to attract foreign investors or borrowed funds for partial financing of projects.

In the nuclear fuel cycle of Canada, which is one of the largest in the world, there is also no monopoly: the most important (but not the only) players in uranium exploration and production are Cameco and Areva, and Cameco and GE-Hitachi compete in the upper levels of nuclear fuel production. Finally, at the beginning of this decade, the state (federal government) refused to monopolize the supply of domestic reactor technologies, and SNC-Lavalin Group, an engineering holding, became the most crucial player in this field.

At the same time, the main equipment market for nuclear power plants is open to foreign suppliers, which occupy an important place in it (in niche manufacturing of steam generators, turbine equipment, etc.) and have the opportunity for further expansion, including the supply of alternative reactor designs to Canadian ones.

Another example is Finland, where the state-controlled Fortum joint-stock company owns a significant portion of some (but not all) nuclear projects, but a substantial part of the assets are held by companies that are not controlled by the government. At the same time, the role of the state in supplying equipment for the nuclear industry is insignificant, as in most other countries that do not have their own key nuclear technologies and production base for the production of the main equipment of nuclear power plants.

Also, competition is developed, and the presence of non-state capital in the nuclear power industry of Spain is significant. However, unlike the United States, Finland, or Canada, there are precedents for preferential control of nuclear power plants by foreign investors (a considerable portion of the assets belongs to the Italian Enel). On a similar liberal path, at least for now, Turkey is moving ahead, allowing competition and full control of foreign investors in its future nuclear industry.

New Market Conditions in the Nuclear Industry

The market conditions for nuclear power differ substantially both in the countries of the group under review and in regions (not necessarily administrative ones) within the same country. The most important characteristics are the degree of liberalization of the electricity market and the working conditions of nuclear power plants on it. In the United States, since the liberalization of some regional electricity markets, all construction or completion projects for new nuclear power units have been carried out in regulated regulatory areas where nuclear power plant owners have the opportunity to enter into long-term contracts for large volumes and to pledge a significant portion of their costs and investments in tariffs approved by state regulatory bodies.

Regional differences and differences in market conditions in Canada's electric power industry are also great, but their impact on nuclear generation is exactly the opposite. For example, the Quebec market has always remained one of the most "regulated" in Canada. In the region, there is a surplus of generating capacities, and mostly cheap hydropower. Against this background, the cost of generating the only "Gentili" NPP operating in this province increased over the years, and in the event of a significant overhaul, which was inevitable, the station would become unprofitable. And the regional authorities decided to abandon nuclear energy; in 2012 the station was closed.

At the same time, in other regions of Canada (Ontario, New Brunswick), where the liberalization of the electricity market has moved further, nuclear generation has survived, although its profitability has suffered from a change in the competitive situation, the need for costly overhauls (for Canadian-designed reactors they are mandatory in about 25–30 years of operation).

The NPPs also fit into the competitive markets of some other states of the group under consideration, where gas generation costs much more than in the United States, and the lack of resources and high costs for reducing greenhouse emissions hinder the development of coal. Under such conditions, nuclear power can remain competitive even against the backdrop of the highest degree of liberalization of the electricity market. The example is Finland, whose nuclear power plants remain highly competitive, despite significant subsidization of unconventional RES, integration of the country into one of the most liberal electricity markets in the world (Scandinavian), a high proportion of cheap hydro generation, developed intersystem network connections in the zone of this market, and so on.

Another difference of this group from the countries of "nuclear absolutism" is the active participation of non-state structures in R&D. For example, in the United States, a significant part of R&D for prospective reactor designs, nuclear fuel cycles, decommissioning is carried out by non-state entities, as well as subsequent introduction of developed industrial technologies. "Privatization" of the R&D sphere reaches such a degree that non-state companies create dual-use technologies. Thus, the industrial technology of separation of uranium isotopes by means of a laser in this country is created by private structures, with the participation of foreign investors—Japanese and Canadian.

For nuclear power in the countries of the group under consideration, the taxation regime common to the rest of the energy sector is usually used, adjusted for the specifics of the industry (the need for deductions for decommissioning and management of SNF and HLW and the like). However, in a number of cases—for example, Spain, which in recent years, has increased the taxation on energy, including nuclear power.

Spanish companies operating nuclear power plants replenished the fund for decommissioning and handling SNF in proportion to the volume of electricity produced—in the approximate range of 0.006–0.008 euro/kWh. In 2013, two taxes were imposed on the handling and storage of spent nuclear fuel, as well as high-level waste. The first is the tax on the unloaded fuel: 2190 euro/kg t. M. discharged from the reactor SNF. The second is the tax on the planned centralized storage of SNF and HLW. It is 70 euro/kg t. M. for SNF and 30,000 euro/m for HLW. Together, both taxes are equivalent to approximately 40–50 million Euros per 1 GW of installed capacity per year.

Two reasons explain tax changes in Spain. The first is Madrid's intention to shift in full the costs of decommissioning and handling SNF and HLW from taxpayers to energy companies. The second is the need to cover the tariff imbalance that arose in the country's power industry due to significant subsidization of RES, which was not compensated by the previously established tariffs. The increase in the fiscal burden in Spain has significantly affected not only nuclear energy, and this distinguishes it from the special nuclear taxes established in some countries with a biased attitude toward nuclear energy.

8

Global Nuclear Investment Environment

In the early 2000s, the nuclear power plant construction world was headed (conditionally, of course) by the Big Four: Americans with their Westinghouse Electric as the flagship of the global nuclear industry, the French with their great powerful AREVA, the Japanese, the Americans with their 104 nuclear power plants, and reorganized into state enterprise ROSATOM.

After Fukushima, two companies became the leading players in the global nuclear industry: from the Russian side, the state corporation ROSATOM and from the US private corporation—the American-Japanese company Westinghouse Electric-Toshiba. Westinghouse, a privately owned company, is, in fact, the world's largest supplier of fuel for light-water reactors, providing 145 nuclear power plants around the world. The company also provides 84% of light-water reactors in North and South America with fuel.

World's Nuclear Investment Environment

Nuclear energy is the basis of the modern economy of the world. In terms of the development of nuclear power plants, the United States is hopelessly

D. Basu and V. W. Miroshnik, *The Political Economy of Nuclear Energy*,
https://doi.org/10.1007/978-3-030-27029-2_8

behind Russia. If 36 reactors are in operation in Russia and 20 are under construction, then in the United States, the last new power plant was built 43 years ago! The renewal of the "park" of power plants in America is almost not performed, and in 10 years they will be forced to close them. At the same time, the United States buys the lion's share of nuclear fuel from Russia. However, the United States does not have the technology for reprocessing spent nuclear fuel, and therefore has to arrange atomic "burial grounds." And the United States hopelessly lagged behind Russia in terms of coverage of the world nuclear power plant construction market. Russia is developing new technologies for miniature nuclear power plants. It has built and operates the world's first floating nuclear power plant.

In the early 2000s, the nuclear power plant construction world was headed (conditionally, of course) by the Big Four: Americans with their Westinghouse Electric as the flagship of the global nuclear industry, the French with their great powerful AREVA, the Japanese, the Americans with their 104 nuclear power plants, and reorganized into state concern ROSATOM.

Back in 1996, Westinghouse Electric sold its defense unit for $3 billion to Northrop Grumman defense corporation, and now Northrop Grumman Electronic System is in charge of US nuclear defense programs. Westinghouse did not stop at that: in the same 1996, all the peaceful parts of the atomic program were sold to the British from the BNFL. The British, in turn, in 2006, sold all the objects of the Toshiba nuclear industry. In this company, the Japanese own 87% of the shares—well, a very controlling stake; another 10% of stocks are owned by Kazakhstan.

After Fukushima, two companies became the leading players in the global nuclear industry: from the Russian side, the state corporation ROSATOM and from the US private corporation—Westinghouse Electric. Westinghouse, a privately owned company, is the world's largest supplier of fuel for light-water reactors, providing 145 nuclear power plants around the world. The company also provides 84% of light-water reactors in North and South America with fuel.

The Nuclear Investment Environment in France

France has the credit to advances in nuclear power by the discovery of radium by Pier and Marie Curie. France depends on nuclear power significantly. The share of the nuclear in total energy production in France is about 80%, but it has the plan to reduce it 50% by 2035 since France has the same problem as the UK, that most of its nuclear plants are old. Decisions are needed whether to invest about $1 trillion to renovate these 58 old reactors or to eliminate its nuclear energy program.

Until recently, compared to coal and gas, nuclear power has been cheaper than any other source of electricity. However, technical improvements in the design of gas turbines mean significant thermal efficiency and less expensive electrical generation from fossil fuels. The sunk cost investment to build a gas-fired thermal station is much less than only one-fourth that required building a nuclear power station. Future oil and gas prices will determine the future of nuclear power developments in France. Nuclear plants can be closed down if the prices of oil and gas are too high. A study prepared by experts, Jean-Michel Charpin, Benjamin Dessus, and René Pellat in the year 2000 indicated these.

The report argues that energy conservation remains the most efficient means to meet future energy targets. Recently, French energy consumption went up. French consumers do not reduce their energy consumption, especially in heating and transportation. France has much fewer options for electricity generation than its industrial partners. Other countries can replace their older coal-fired power stations by new nuclear power stations. In some states, like in Sweden and Germany, public hostility regarding radioactive waste forced their governments to close their nuclear plants, the French government does not have this option, due to several reasons.

France's nuclear weapons development began on February 13, 1960, with the country's first nuclear weapons test. France collaborated with Israel to develop nuclear research and weapons development. As of 2014, the Arms Control Association estimated that France had about 300 nuclear warheads, most of which were submarine-based. In February 2015 Francois Hollande suggested other countries should follow France and stop the production of fissile materials and nuclear tests, which can be replaced by simulation tools (Bruno Tertrais in www.frstrategie.org, 2019). In early 2019 France has exploded a nuclear device through simulation techniques.

The Nuclear Investment Environment in Sweden

In this country, energy companies have been paying a tax on the installed capacity of nuclear power plants since 2000. In the event of power unit downtime, there is a discount, but it is not large: about 3% of the tax amount. The authorities several times increased the rate of the nuclear tax, besides it grew at the expense of increase of thermal capacity of some power units within the limits of programs of their modernization. Estimates of the specific weight of the tax in spending on nuclear generation vary considerably, but in all cases, they are measured by tens of percent. In addition to the power tax, owners of Swedish nuclear power plants pay high contributions to SNF management and decommissioning funds from specific industry charges.

Discriminatory taxation and protectionism concerning renewable energy artificially reduce the competitiveness of the Swedish nuclear power industry. So, according to Vattenfall, the tax on the installed capacity of nuclear power plants was one of the main reasons for the record net loss of 19.8 billion Kroons (2.1 billion Euros) incurred by the company in 2015. The owners of the nuclear power plant tried to challenge the tax in the Swedish courts and the European Court, but without success.

It is no accident that the owners of some nuclear power units announced their imminent closure, in the absence of approved replacement plans. Until 2020, it is expected to finally stop four of the ten nuclear power units in operation in Sweden (30% of the country's nuclear capacity), one of which (No. 2 of the Oskarshamn nuclear power plant), previously stopped for modernization, official decommissioning. Two controlling owners of Swedish nuclear power plants (Vattenfall and E.ON) justify the need for early closure of a part of the facilities with low electricity prices, increased costs and the negative impact of the tax on nuclear generation.

The investment climate in the Swedish nuclear industry is also affected by the uncertainty of the country's energy strategy, backed up by historical experience. In the 1980s, Swedish authorities planned to abandon nuclear generation by the beginning of the current century but later abolished this proposal, confining themselves to a selective shutdown of some capacities (the two-unit NPP "Basebek" was finally shut down). However, after the new ruling coalition came to power in 2014, the issue of the gradual abandonment of nuclear energy in favor of renewable energy is once again raised at the political level. All these fluctuations in the energy strategy rob the investors of nuclear power of confidence in the future and prevent planning long-term investments.

Greenhouse emissions in Sweden are more expensive than the EU emission limit for exceeding emissions quotas (100 euro/tonne CO_2). However, the taxation of greenhouse emissions almost does not concern the electric power industry. Electricity generation on fossil fuel is an insignificant part of the country's generation balance (~ 3%), with condensation plants and CHPs paying for emissions only from a small fraction (1.5–5% maximum) of the fossil fuels they consume. Thus, the factor of fiscal discouragement of thermal generation competing with nuclear power plants is practically irrelevant, unlike the example of Great Britain and several other countries described above. At the same time, distributed renewable energy sources (RES) receive significant preferences, primarily within the framework of a system of "green" certificates, which are ultimately paid for by electricity consumers. Against this background, the role of distributed RES in the energy balance is steadily growing: in particular, in 2015 their output increased by 45% and amounted to about 17 billion kWh per year (11% of Sweden's generation).

The Nuclear Investment Environment in the Czech Republic and Finland

The construction of the Temelin NPP began in 1985. By 1990, it was decided to suspend its construction in connection with the "insufficient reliability of the Soviet bloc's." However, the shortage of electricity forced the Czech leadership to review the strategy—it was decided to finish building two units using the developments of the company Westinghouse.

The first reactor was launched in 2000 and the second reactor in 2002. Both blocks were commissioned in 2002 and 2003, and in 2007 problems began. The fact is that from the very first loading of fuel into the reactors, deliveries were made not by the Russian company TVEL, but at that time by British, recent Americans (Westinghouse Electric Company) and future Japanese.

The Temelin NPP has become an experimental ground for creating an alternative to Russian fuel for VVER-1000 reactors. After several years of operation, the deformation of fuel assemblies (fuel assemblies) was revealed. A little earlier, experiments began in Finland on VVER-440 units. In 1998, Westinghouse first attempted to master the fabrication of fuel cassettes for these reactors. At that time, five fuel assemblies for the experimental operation were delivered to the Loviisa nuclear power plant. They were loaded together with six Russian fuel assemblies. In 2002, all 11 cassettes were unloaded and investigated.

The corrosion resistance of Westinghouse cassettes was significantly lower than the Russian ones. The corrosion layer in American assemblies was 21 μm, while for Russian meetings it was only 1–4 μm. Also, "American TVELs" recorded the release of gaseous fission products, twice the equivalent of Russian fuel assemblies. As a result, the owner of the station, Fortum, decided to cease cooperation with Westinghouse, and so far no experiments have been done on the VVER-440 units.

Even before the accident at the Temelin NPP in August 2005, the first block of fuel from Westinghouse was loaded in the third unit of SUNPP instead of traditional fuel from TVEL. This program was agreed upon under Kuchma, when in 1999 Westinghouse co-founded the Center for Designing Active Zones based on the National Science Center "Kharkov Institute of Physics and Technology." This decision was the beginning of the project "Nuclear Fuel Qualifications for Ukraine" as part of the agreement between Ukraine and the United States in 2000 to diversify nuclear fuel supplies.

In 2012, in the seventh year of fuel exploitation, its deformations were noticed. In the eighth year, they became critical, and US fuel assemblies were hardly extracted from the reactor. By the way, in the Czech Republic, the problems were also revealed in the seventh year of fuel exploitation, which is not accidental. The processes in the reactors are the same. In general, such stability of the emergence of problems shows that they are not random.

After this incident, cooperation with the United States was curtailed, despite the conclusion in 2008 of an agreement to transfer a minimum of three (six) Ukrainian VVER-1000 to US fuel in 2011–2015. The fact is that the NPP market is not flexible enough, but it has its tendencies. Westinghouse supplied its fuel for seven blocks in Belgium, one block in the Netherlands, two blocks in Finland and two blocks in South Africa (all in half with AREVA), 14 blocks in the UK, one in Slovenia, 48 in the United States, two in Taiwan, two in Switzerland, and seven blocks in Sweden.

At the same time, the Belgian nuclear power plants will be closed by 2025, and the process of stopping the blocks will begin very shortly. All the old British reactors are likely to be shut down in the 2020s, and instead, Europeans, led by the French, will build new ones. Moreover, then Westinghouse has no chances to get contracts.

In Sweden, in 2014, the "green" came to power, which adopted a program to replace nuclear power plants with clean energy, following the example of Germany. They raised "atomic taxes," which force the nuclear workers to close their stations ahead of schedule. In 2015, the owners announced an early stop of the four blocks. In 2016, NPP operators made a statement that if the government's policy does not change, they will close the remaining blocks in the 2020s.

In Taiwan, there is still debate about the future of the nuclear industry, and if the opponents of the nuclear power plant win, Westinghouse will lose two more blocks. No less significant losses are waiting for the corporation in the American market. Only in 2013, four blocks from the Westinghouse pool were closed at once. The shale gas signed the verdict. The work of the stations became unprofitable, and therefore, even having a license for work, the operators were forced not to extend the service life of the stations, but to close them to avoid financial losses.

Westinghouse had the problem of survival. Ultimately it could not survive. Global Nuclear Fuel-Americas (GNF-A, a subsidiary of GE-Hitachi) signed a cooperation agreement, which aims to organize joint activities on licensing, marketing and fabrication of fuel for operators of light-water reactors (PWR) in the United States. Under the terms of the agreement

JSC, TVEL, and GNF-A will form a strategic alliance to offer customers fuel assemblies based on the Russian design of TVS-SQUARE for pilot operation, licensing and obtaining permission from the US Nuclear Regulatory Commission for the commercial supply of this fuel. Fuel assembly "TVS-Square" in the future is planned to be produced at the GNF-A plant in Wilmington, North Carolina. GNF, as we recall, is the largest supplier of nuclear fuel at Japan's nuclear power plants—22 units. The company also operates in the US market—27 blocks, Mexico—2 blocks, and Switzerland—2 blocks.

The creation of a strategic alliance for the promotion of Russian nuclear fuel to the United States, Mexico and, probably, in the event of the resuscitation of the country's nuclear program in the Japanese market, is a powerful blow to ROSATOM's competitors. Just ten years ago, Westinghouse was the main competitor of ROSATOM and threatened to close the commercial nuclear fuel market outside of Russia, and today, everything has changed places. Westinghouse retreated in all directions, clinging to Ukraine as the last saving straw, and ROSATOM not only contracts for a contract for the construction of new nuclear power units but also makes an attempt to oust Americans from their domestic market.

The latest agreements between GNF and ROSATOM, the sale of Westinghouse first to the British, and then to the Japanese say that the United States lost the war on the nuclear front. Everything that happened further, since the collapse of the USSR, is an attempt to turn history back. That was an unsuccessful attempt because it was based not on economic justification, but political pressure. On the contrary, and this can be said confidently, the events of the last decade show that the nuclear future belongs to companies, none of which is American.

The Nuclear Investment Environment in the United Kingdom

In the United Kingdom, the state refused to own or manage most of the assets of civilian nuclear power. The participation of authorities in the industry is manifested not in the form of a total state monopoly, but in the creation of privileged regimes for state regulation of business in the nuclear field with the aim of almost wholly replacing the nuclear park with new blocks and further increasing it. A characteristic feature of the investment climate is the continued commitment of the state to the further development of nuclear power, despite government changes and the adverse impact of the Fukushima accident on public opinion, which, however, remained mostly supportive of atomic energy, unlike many other states.

The future development of nuclear generation in the UK is planned primarily through foreign investment and technology; at the same time, the state is ready to compensate investors for a fair share of risks. Recently China purchased a nuclear power plant that was about to be closed down.

Measures to encourage investment in nuclear energy include a privileged long-term pricing system for new nuclear generation projects, government financial guarantees for some projects, and stringent measures to discourage greenhouse emissions, which generally reduce the competitiveness of electricity production on fossil fuels.

A unique system of electricity pricing for some generation facilities planned for construction provides for the conclusion of individual long-term contracts (for several tens of years) with investors such as so-called contracts for a difference, providing for them a return on investment in the construction of a station and an established level of profitability, irrespective of changes in prices for electricity market. This pricing system is based on the so-called feed-in tariffs model, which in one form or another, is used in many other countries, mainly concerning distributed RES. The first nuclear power generation project to be implemented using this scheme should be the construction of two power units at the Hinckley Point NPP, planned jointly by the French EDF and the Chinese CGN. The investment decision on this project is expected until the end of spring. If these investors implement subsequent projects (the construction of power units at the nuclear power plants "Sizewell" and "Bradwell"), the guaranteed price level of the contract will decrease somewhat due to the expected cost reduction.

The next "gift" to investors in nuclear energy can be the provision of government financial guarantees, provided by the government for significant infrastructure projects, not only in the nuclear field. Another measure that contributes to the development of nuclear power in the United Kingdom is increased taxation of emission energy sources, primarily thermal generation on organic fuels. As in other EU countries, the UK provides for a fee for the emission of greenhouse gases, including for electricity producers. However, in recent years, London has significantly increased the cost of such emissions, setting a premium to the price determined within the European Emissions Trading System (ETS). The most significant pressure is experienced by coal-fired power plants that are not equipped with emission control systems—direct competitors of the NPP in the base generation niche: emissions are more expensive, also, until 2025 such stations are planned to be closed.

Unlike several other countries, the UK tries not to overload its nuclear power industry with a tax burden, treating it as a priority area. As a result, the United Kingdom today is leading in Europe in terms of the number of planned nuclear power unit construction projects and participating investors. It is designed to build up to 18 GW of new nuclear capacity.

Despite the significant deterioration in recent years of market conditions, Britain managed to attract several foreign investors to nuclear projects: three consortia formed by the largest energy companies and technology suppliers.

Taxation of Nuclear Generation

For all the differences in taxation objects, a common feature is that the tax burden is not directly related to the volume of electricity production and is charged almost independently of it, so the real level of fiscal pressure on nuclear generation in these countries has increased over time—as at the time of rising tax rates, and in periods of unchanged rates. For example, in Germany in recent years, the development of nuclear power plants has declined (in particular, the average CIUM on the Isar, Gronde, Nekarvestheim, Philipsburg units was reduced), but this led to an additional increase in the tax burden per 1 kWh of output.

Nuclear Discrimination

For several indicators, countries with an unfavorable investment climate are close to some states from the other two groups. For example, in their nuclear power, as a rule, there is no state monopoly, and the role of atomic energy in several such countries is very high (up to 40–55% of the generation balance). However, in the countries of the group under review, the state establishes discriminatory conditions for the industry that lead to its reduction and gradual elimination or impede development, sometimes despite favorable economic prerequisites: significant demand for inexpensive energy and high dependence of states on energy imports, and even electricity.

This group identifies countries and regions that have set themselves, on a legislative or normative level, the goal of renouncing nuclear power in specific or flexible terms (Germany, Belgium, Switzerland, Taiwan, Scotland) or instituting an actual ban on the use of atomic energy for generation (Italy, Austria and others). So, Germany plans to abandon nuclear age in 2022, Belgium and Taiwan—in 2025, Switzerland—presumably in 2034.

Under such circumstances, long-term investments in the development of nuclear generation, lose their meaning. In some of these countries, the situation is further complicated by the discriminatory fiscal regime for the existing nuclear power industry: individual taxation, as well as excessive state support for competitive generation activities. Against the backdrop of the deterioration of the situation in many electricity markets since the end of the 2000s, the profitability of nuclear generation would be reduced.

An example is Germany, where, a nuclear fuel tax of 145 Euros was charged per gram of fissile material being loaded into the reactor—235U, 239Pu, 241Pu, and 233U (in Germany, MOX fuel and fuel from regenerated uranium). It should be noted that Germany's nuclear power industry is generally characterized by higher average profitability before taxation than for several other countries. Unlike many states, wherein conditions of deteriorating market conditions, part of nuclear power plants with relatively old nuclear blocks of medium capacity are balancing on the verge of profitability and are sometimes closed by owners for economic reasons (USA, Sweden, Canada, Belgium, and Japan), only highly effective nuclear generation on relatively modern power units.

Despite the potentially higher stability of German NPP operators toward discriminative fiscal measures, power companies are forced to reduce their nuclear business, up to the closure of nuclear power plants ahead of the deadline approved by the authorities of Germany. So, in June 2015, for this reason, the Grafenreinfeld nuclear power plant was terminated prematurely.

In Germany in recent years, the development of nuclear power plants has declined (in particular, the average CIUM on the Isar, Gronde, Nekarvestheim, Philipsburg units was reduced), but this led to an additional increase in the tax burden per 1 kWh of output.

All owners of German nuclear power stations demanded financial compensation from the state in courts of different instances in connection with the early termination of nuclear generation by 2022 and the preservation of the tax on nuclear fuel (which was introduced in 2010, on the contrary, the extension of the nuclear power plant operation). Although on several claims, energy companies have achieved moderate success (the Tax Court in Hamburg declared illegal the preservation of the tax on nuclear fuel), the remaining authorities (the European Court, the Federal Constitutional Court of Germany) have not yet supported the requirements to the authorities.

Another example of such discrimination is Belgium, where in 2008 a tax was imposed on nuclear generation in the form of an annual fixed payment of owners of nuclear power plants, which gradually increased. Belgium's state regulators do not deny that the tax is comparable to the profits of NPP owners from all activities in the energy sector in Belgium, which has declined several times in recent years due to the worsening of the market situation and the growth of unit costs for nuclear generation (at a number of Belgian power units in the current decade there were extended unplanned downtime associated with identified defects and breakdowns).

According to the owners themselves, in recent years, this business is unprofitable, and above all because of the tax on nuclear generation. In mid-2013, co-owners of Belgian nuclear power plants (subsidiaries of French energy companies GDF Suez and EDF) initiated an action in the Constitutional Court of Belgium, disputing the increase of this tax. However, the court rejected him and the subsequent appeal, recognizing the efforts of the authorities as legitimate.

A similar discriminative fiscal regime for nuclear power plants, combined with preferences for competing for generation, is also applied in several countries that in the coming years do not intend to abandon nuclear energy or are going to consider this possibility only in the uncertain future. A typical example is Sweden, whose government plans to maintain nuclear power for a long time—until it becomes possible to replace it with an energy balance (35–50% of generation, depending on significant fluctuations in hydro generation) by RES.

For all the differences in taxation objects, a common feature is that the tax burden is not directly related to the volume of electricity production and is charged almost independently of it, so the real level of fiscal pressure on nuclear generation in these countries has increased over time—as at the time of rising tax rates, and in periods of unchanged rates.

Operated Nuclear Power Plants

The first mention of the danger of using nickel alloys for the production of equipment of Western nuclear power plants was made in the work of Coriou (1959).

The safest reactors in the world can be closed because of their increased danger. Independent experts who investigated the insurance case at the Davis Bess nuclear power plant issued a conclusion on the suitability of stopping all 69 units with PWR reactors in the United States. The regulatory body of the United States does not hide its anger. However, over time, everything has calmed down, and the process of extending the operation of the nuclear power plant continues because, with the new NPPs that are being built with AP-1000 reactors, problems seem even more. The alloy 600 was actively used earlier in the western projects of light-water plants. This material eventually created one of the biggest crises in the history of reactor construction.

In 1971, the first leaks in the PWR steam generators caused by the corrosion of alloy 600 were recorded. Then, on the block No. 3 of the San Onofre NPP with the PWR reactor, a nozzle for the instrumentation on the volume compensator was corroded. In 1987, the sleeves of the same compensator flowed on the block with PWR in Arkansas. Moreover, in 1989, the first information appeared on the corrosion of PWSCC on French blocks. Two years later, the first cases of corrosion damage to the nozzles for the entry of CPS drive on the lids of Western reactors under pressure were discovered. Today, there is leakage of measuring sensors in the bottoms of reactors and the appearance of cracks in the butt welds. Well, it all started as much in 1971 with a leak in the steam generators.

Anyone who tries to understand the problem of alloy 600 independently is critical to know that the main difficulty is that this material has been used too often by Western designers in various elements of the power unit. It is worth looking at the diagrams of major Western projects to see the unusually wide prevalence of this alloy. Corrosion PWSCC has become an integral companion of Western reactors. All materials of the primary circuit will degrade. This pessimistic conclusion was made in 2004 by specialists from the Institute of Nuclear Energy (NEI).

Nuclear Power Plants (NPPs) Under Construction Worldwide

For the first time since 1979, on February 9, 2012, the US Nuclear Regulatory Commission (Nuclear Regulatory Commission) decided to issue a "combined license" for the construction and operation of two AP-1000 reactors at the Vogtle NPP in Georgia. However, before building a new pressure cooker at home, the United States decided to first break in technology in China, which has long been striving for atomic superpower status. China is overshadowed by a sense of shame and helplessness arising from the fact of progress in the creation of new reactor projects. The republic still does not have its single brand among commercial power reactors.

The 11 units operating in China use different technologies. In Guangdong, there are reactors from France, in Tianwan—from Russia, and on the third stage of the Qinshan NPP—from Canada. Only at the first Qingshan block is a reactor that could be called Chinese, but it naturally belongs to the second generation.

The Qinshan-1 block is a subject of historical pride of the Chinese. The only unpaired unit in the PRC, it is also the first Chinese nuclear power unit, was created by the Chinese enterprises and departments. The Chinese are using this project for deliveries to Pakistan—the only country that buys power reactors from China.

At the end of 2006, the Chinese authorities decided that it was time to "unify the technological roadmap for the development of nuclear energy." Behind these words lies the next strategy—to acquire technology, understand technology, localize technology, and start innovations based on it. China plans to focus on work with the AP-1000, which is here called the most developed project of the third generation.

In May 2007, the company SNTPC was created, and it was entrusted to implement, using the example of the AP-1000, a strategy for unifying the roadmap. As soon as China chose in favor of the AP-1000 at the end of 2006, many wondered why the United States does not build units with such reactors if they are as safe and advanced as the advertising says?

There were doubts, but the PRC decided to head the direction of the AP-1000 because it considered that it would save her time and money. One could wait for the appearance of other customers at the AP-1000, the reference blocks, and only then decide on their order for China. HE But in this case, the country would have to pay for the blocks much more expensive, and it would not have happened soon. As it became clear later on, the Americans did not neglect the AP-1000 technology for themselves. The first AP-1000 nuclear power plant will be commissioned in the United States in 2016, or just three years later than in China.

Naturally, the Chinese have encountered unforeseen difficulties and problems at almost every stage of construction, and the Americans, considering the Chinese experience, have begun their construction. In the United States, Westinghouse begins the development of an AP-1000 NPP, nicknamed "the radioactive tube."

Reference

Coriou, H. (1959). *Colloque de Metallurgie sur la Corrosion* (p. 161). Amsterdam: North Holland Publishing.

Part III

Prospect of Nuclear Power Development

9

Germany as Nuclear Power

Nuclear Secrets of the Third Reich

Research on nuclear energy and weapons have started in Germany in the early 1930s. In this chapter, we trace those developments. We begin with a report of Major-General V. A. Kravchenko, Deputy Commissar of Internal Affairs of the USSR, written in 1945 on the work done and the identification of data on research in the field of nuclear physics in Germany:

November 12, 1945

Top secret

In accordance with the tasks assigned, a group of operative and scientific workers examined the following research institutions and enterprises that worked with Problem No. 1: Leipzig, Halle, Jena, laboratories in Kummersdorf, Ronneburg, Oberschlem, Vaide, Vahe, the Freiberg mountain academy, Koch and Shterzel in Dresden, Lane in Merseburg, Auer and Oranienburg, plants "Zeise" in Jena, research centers in Peenemünde, Nordhausen and elsewhere.

The researchers who worked in the sectors related to Problem No. 1: Bongofer, Berkai, Herman, Chulios, Bori, Pose, Weiss, Gabe, Hund, Fisher, Zimmer, Ive,

D. Basu and V. W. Miroshnik, *The Political Economy of Nuclear Energy*,
https://doi.org/10.1007/978-3-030-27029-2_9

> *Westmeyer and others were identified and interviewed. As a result of the study of the collected materials, organizations and scientists engaged in solving specific issues related to Problem No. 1 were determined, and a scheme for organizing work in Germany on nuclear physics was drawn up.*
>
> *All this allowed the group to establish the scope of work on problem number 1 conducted by the German scientists: the final results of individual research projects; determine the number of scientific research organizations and scientists, establish the approximate amount of uranium and heavy water that Germany had before surrender; to establish also the quantity of uranium and heavy water exported by the Allied forces. At the same time, enterprises and scientific research organizations planned for export to the Union.* (https://web.archive.org/web/20180624105235, http://www.alexanderyakovlev.org/fond/issues-doc/7312)

The systematic bombing of cities in which the above-mentioned laboratories and factories were located, as well as the rapid advance of the Red Army and allies, placed the German research organizations dealing with Problem No. 1 in a complicated situation and, according to reports of German scientists, did not allow the Hitler-government to systematically conduct evacuation of research organizations in safe places. For this reason, much of the equipment and technical documentation was destroyed. Most of the German scientists along with their materials, and literature were with the Allies, only some of the scientists remained in the territory occupied by the Soviet troops. Bombing and fires destroyed almost all buildings of laboratories; their property was subjected to multiple attacks.

Until 1939, work on uranium in Germany was carried out by many scientists, but this work was theoretical. Since 1940, several physicists like Khan, Strassman, Geiseiberg, Flyuge, and others have established the possibility of obtaining in large quantities of intra-atomic energy. A special sub-department of nuclear physics was created in the German Armaments Department. The head of the research department of the Armaments Department was Professor Shuman. Dr. Dibner was appointed the head of the nuclear physics sub-department.

To test the main issues related to the military application of nuclear physics, Dibner organized a specialized laboratory at the military training ground in Kummersdorf (near Berlin), which was directly part of the

Armaments Department and was thus an army organization. The scientists selected of this laboratory were Chulios, Herman, Gartwick, Berkei, Hülsman, and other members of the Nazi Party. Before the military laboratory in Kummersdorf, the task was to use the work carried out in all the institutes in Germany to create a uranium machine suitable for use as an engine for missiles and submarines and the development of an atomic bomb.

Simultaneously with the creation of the military laboratory in Kummersdorf, Dibner collected all the work on the uranium problem of several major German scientists who were working on solving individual problems. Later, by the end of 1944 approached the final solution to the problem of creating a uranium machine.

It was established on the basis of individual testimony of scientists that the Hitler government until 1943 did not attach particular importance to work on uranium for two reasons: firstly, obviously because these works were of a theoretical nature and seemed to be far from technical implementation and secondly, because the government attached much more importance to the work on the more real and practical, the creation of rockets of the Fau type, Mallett, radio-controlled torpedoes.

The latter is confirmed by the enormous effort of the construction of exceptional research grounds, laboratories, and plants for testing and production of jet aircraft and shells. During the war, the Germans in Peipemupde carried out extensive work on the creation of test rigs for Fau, built industrial premises—ground and underground, living quarters for workers. Dozens of cooperative plants were created.

Hundreds of thousands of people were involved in the work on creating bombs and guided bombs. At the same time, before 1943, work on the uranium problem was carried out by disparate groups of scientists whose work was not purposeful, and some scientists, judging by their testimony, were skeptical about the possibility of use of intra-atomic energy for military purposes.

In early 1943, all work on the uranium problem was transferred to the State Research Council (Reichsforhsgsrat), led by Goering. Goring appointed a well-known German physicist, a member of the NSDAP, Professor Ezau, a specialist in high frequency, the head and organizer of the work on the remote control of rockets, torpedoes, and also in the field of personal radio communications for uranium work and nuclear physics.

Due to the fact that the leadership of Ezau did not produce the proper results and, besides, he was not a specialist in nuclear physics, Ezau was replaced in late 1943 by a well-known expert in this field, Professor Gerlach, who until the end of the war remained the authorized representative of Goering on questions nuclear physics, while also being the head of the Laboratory for Atomic Physics at the University of Munich. The immediate organizer and leader of all work on the uranium problem were Dr. Dibner, who remained in place, despite all the movements of the top management—Schumann, Ezau, Gerlach.

Since 1943, the entire organization of work on the uranium problem has acquired a distinctive character. The main center, which was supposed to use for almost military purposes the results of all the actions, was the Dibner Laboratory in Kummersdorf. For this purpose, Dibner gradually attracted the most critical scientists for setting up the development of individual problems and discussing various issues resolved in Kummersdorf: Heisenberg, Dopel, Bothe, and others. Thus, work was carried out in most of the most significant theoretical centers in Germany, i.e., in the Kaiser Wilhelm institutes: for physicists—Professor Heisenberg, for chemistry—Professor Khan, for Medicine—Professor Bothe, for Physicochemical Institute in Hamburg—Professor Garteck, Physicochemical Institute, University of Munich—Professor Gerlach and Kluzius, Physical Institute, University of Leipzig—Professor Hoffmann, the Physics Institute of Cologne University—Professor Kirchner and others.

Along with the leading scientific centers, scientists from other institutes and educational institutions were also involved in the work of Dibner, in particular, Professor Enzen—Higher Technical School in Hanover, Professor Ritzler—University in Boyne, Professor Kopfermap—University in Göttingen and many others. A total of about 86 different scientific institutions and over 100 people were involved in the work on the uranium.

Simultaneously with the formulation of scientific tasks in the above-mentioned leading scientific organizations, the Directorate of the special commissioner for nuclear physics was involved in manufacturing plants for processing and obtaining metal uranium—the plants "Auer" and DGUSSA, the last in this respect was directed by Dr. Riehl, as well as chemical plants for the production of heavy water. A special place in this system was occupied by the radium institute of the State Physico-technical Institute, headed by doctors Boite and Weiss, who was engaged in the manufacture of measuring apparatus and neutron preparations and stored all the German radium reserves.

Also, the Kaiser Wilhelm Institute of Biophysics in Frankfurt am Main, with offices in Bach and Oberschlem, headed by Professor Raevsky and Dr. Crevs, focused on the development of biophysical problems related to atomic physics. The official subject of this institute was to work on protecting the human body from radium and creating artificial radioactive substances. Work on the uranium problem was carried out in the following main area: Theoretical and experimental work, the purpose of which was the creation of an operating uranium machine.

The main works in this direction were conducted in the laboratory in Kummersdorf—with the head Dibner, and in the laboratories of the Kaiser Wilhelm Institute—with the head Geiseberg. In both laboratories, laboratory models of uranium machines were created, and they were thwarted by the evacuation that began as a result of intensive bombardments. Thus, the Kummersdorf laboratory, leaving Kummersdorf on the challenging Stadtilm-Innsbruck route, could not even begin to work, having lost all equipment and all people. The laboratory of the Kaiser Wilhelm Institute, based after evacuation in Hechingen (the zone of occupation of the Allies), according to available data, achieved only at the end of the war the creation of a laboratory model of a uranium machine. During the evacuation of the laboratories of the Kaiser Wilhelm Institute, Professor Pose, who was commissioned to create his base for work on the uranium machine at the Institute-Leipzig-Halle, did not have time to deploy his work.

Research and Experimental Work on the Production of Heavy Water

The main work on obtaining heavy water was carried out in Norway with the participation of Professor Gartek. Considering the need for getting heavy water in Germany itself, Professor Gerlach put a series of experiments on obtaining heavy water with the involvement of Bongofer, Gabe,

Herald, Hartek, Virts, and Sius. Pilot plants for heavy water production were created at the Leyna plant in Bitterfeld, but no positive results could be obtained until the end of the war, in particular, the large installation at the Leyna plant was destroyed by air bombardment.

Separation of Isotopes

That was the main question on which the possibility of making an atomic bomb depended, Professor Gartek, Dr. Bage, Professor Bothe, Dr. Jenschs in Vienna, and others were doing. Professor Gartek achieved the most successful results in obtaining significant quantities of uranium-235 in Hamburg with the ultra-centrifuge he invented. This device was adopted for serial production at the Anshutz factories in Kiel, but these works were also delayed due to bombardment and evacuation and, according to reports, were not completed before the end of the war.

Thus, on the basis of all the above data, it can be stated that, in spite of the considerable amount of theoretical and experimental work in the course of which definitely sure German scientists Gaizsnbsgr and Gartsk prepared a solution of almost all the main issues needed to resolve the uranium problem, this problem of an active uranium machine or atomic bomb was not solved until the end of the war.

According to the group of scientists Bongofsr, Hund, Pose, Weiss, Chulios, Herman, and several others in Germany, there was a complete opportunity for the successful completion of the work on the creation of a uranium machine and an atomic bomb. In Germany, for example, there were about 10 tons of uranium metal with a raw material stock of about 500 tons of pure uranium oxide in 1944, about 2 tons of heavy water from Norway, but practical results were not achieved, according to these scientists, mainly because German specialists who dealt with the uranium problem avoided making specific commitments to build atomic devices suitable for military purposes. The same caution was also shown by the official leaders responsible for setting up the uranium problem in Germany—Gerlach, and Dibner. An attempt by the formal leaders of the work on nuclear physics in Germany, in particular, Dibner, to control the work of non-partisan specialists and to use their achievements with the help of the Kummersdorf military laboratory did not lead to proper results, mainly because this laboratory did not have a sufficient number of qualified physicists, which could competently evaluate the achievements of individual scientists who have successfully solved specific issues related to nuclear physics.

As a result of the defeat of Germany, the most significant scientific workers, in particular, Gaizsnbrsr, Khan, Flygs, Kluzius, Gartek, Gerlach, Dibner, Virts, Strassmann, Bage, Bote and others, were in the occupation zone of the Allies, and all the materials scattered in result of repeated evacuation. On the territory occupied by Soviet searches, currently identified by scientists associated with the problem No. 1 were: Professor Bopgofer, Dr. Weiss, Professor Pose, Dr. Rexer, Dr. Herman, Dr. Chulios, Dr. Fisher, Dr. Westmeyer, Dr. Berkei, Professor Hund, Dr. Gabe. Given that the overwhelming majority of workers in this laboratory were members of the Nazi party, it would have been necessary to create special conditions for this group, corresponding to closed-type laboratories. In the process of working on questions on the problem No. 1 by the group, there were found in Germany the existence of severe scientific organizations dealing with the issues of the connection of atomic physics with biology and physiology, which had a military purpose. In particular, for these works during the war, installation of an ultra-high voltage of 3 million volts, discovered in disassembled form in salt mines in the town of Stasfurt, was built.

The leading institute dealing with biophysics is in the zone of the American occupation in Frankfurt am Main and in Ockstet. In the Soviet area, there is one branch of this institute in Oberstlem and a laboratory in Waha, as well as a brain institute. A survey of these enterprises and a survey of their employees by Crevs, Zimmer, and Born showed that the institute was engaged in studies on the effect of radioactivity, X-rays and neutron bombardment on living organisms, and in particular on bacteria. During the inspection of the laboratory in Oberstlem, the workers of the group seized military bacterial preparations that were in the laboratory for studying the effect of these emissions on the growth and mutation of organisms.

It can be assumed that this installation of 3 million volts was intended for a repetition of these experiments on a large scale since similar preliminary tests were carried out at a facility of 300,000 volts. Given the availability of scientists working in this field—Crevs, Timofeev, Zimmer, Born, as well as the availability of laboratory equipment in Oberstlem and Vahe and the installation of a high voltage of 3 million volts, it was expedient to create in the USSR a unique closed laboratory, which must deal with the study of special effects of radioactive emissions on living organisms.

The group finds it useful to export the atomic physics laboratory in Ronneburg, the laboratory for studying radioactivity in Oberstlem and the branch of the Institute of Biophysics in Vahe (http://imhotype.livejournal.com/221198.html). Based on the above, there was a decision to do the following:

Instruct employees of the 1st Main Directorate of the SNK of the USSR based on all materials collected by the group, as well as articles on the state of the uranium problem in Germany, available in the 1st Main Directorate, to conduct a full analysis of the state of this problem in Germany. To instruct the specialists and operative workers to study in detail the feasibility of creating two closed-type laboratories.

Many aspects of the Second World War will remain "little-known facts" if only because in 1946 the Rockefeller Foundation allocated 139,000 dollars to write the official version of the tragic events, thanks to which the cooperation of the Third Reich and Standard Oil still belongs to "little-known facts."

After the war, this company was the owner of several patents, Otto Ambros, who participated in the development of poisonous substances called sarin and soman. Ambros was the Fuhrer of the military economy, combining the leadership of the production of rubber and toxic gases in IG Farben with the direction of the Committee on Combustible Chemicals in the Ministry of Armaments, as well as the plants in Auschwitz, Shkopau, Ludwigshafen, and Opau.

To work in a well-known German concern, which was the locomotive of the economy of the Third Reich, Ambros joined in 1926, when he was sent to Sumatra to study natural rubber in order to reproduce his synthetic analog, and by 1935 he had become a leading expert in the field of artificial rubber, Under his leadership, the Auschwitz-III technological lines were built, which was planned as a grandiose rubber plant.

Without a doubt such a person could not help becoming a "useful Nazi," a term that was put into circulation in the United States back in 1944, and the "allies" tried to prevent such Nazis from falling into the hands of Soviet troops. For example, 45 chemists Loyne from IG Farben were removed from their homes in the Russian zone to the west.

In search of "useful Nazi" in the field of chemistry, Colonel of the British chemical service, Paul Tarr, interrogated Gerhard Schroeder in the factory laboratory of Elberfeld. He had no idea about the whereabouts of one of the leading technologists IG Farben.

At that time, Ambros was already discovered by American "Nazi hunters" in Bavarian Gendorf, where, at the entrance to the city, they met a man who introduced himself as an ordinary chemist and showed American soldiers an underground factory for producing soap and cleaning products for the domestic market, hidden underground from bombing allies.

The chemist was detained and, just in case, had to be sent to Aschan filtration center in Luxembourg. When the Soviet troops got a chemical weapons warehouse in the Silesian Duhernfurth, Lieutenant Colonel Tarr sent an escort to the detained Ambros to bring the war criminal as an expert, but Ambros, along with the companion, went to the French control zone, where he remained as an employee of the plant BASF in Ludwigshafen.

Strange war criminal, moving between the zones of the Allies at his discretion, even accompanied by an escort, still appeared before the judges of the Nuremberg trial. However, before it began, US Army lieutenant Walter Jessel has already identified loyal "US scientists from Germany" who were subjected to early release on the lists of JIOA (Joint Intelligence Objectives Agency). After such order by John McCloy, the US High Commissioner in Germany, Ambros sets to work as an "adviser" at Dow Chemical, as well as in the chemical armies of the US Army, and the US Department of Energy.

In 1951 Peter Grace, a close friend of Ronald Reagan, provided Otto Ambros with a leading position in his company W.R. Grace Company. The company in which he worked after his release without a doubt can be attributed to the structures associated with the Third Reich. The American headquarters of Thyssen Inc., the leading NSDAP financier Fritz Thyssen, was located in the W.R. building in New York.

In 1978, 29.6% of W.R. Grace Company for four hundred million dollars purchased Flick Group, combining on the European continent its turnover of 3.2 billion dollars with a 5 billion new subsidiary. Friedrich Flick, who was a prominent sponsor of the NSDAP, was also the most significant philanthropist of the SA and the SS and a friend of Himmler and Goering. Friedrich Flick was the most significant industrial figure of the Third Reich; he was a member of the Presidium of the Reichsvereinigung Eisen cartels and the Reichsvereinigung Kohle (Imperial Coal Association), the council for the industrial association (Virtschaftsgruppe) of the steel industry Eisen schafffende Industrie, in particular at the plant in Hennigsdorf. Due to apparent secret agreements, the Himmler supervising the concentration camp system, with the assistance of one of the directors of Vereinigte Stahlwerke Steinbrink, received 100,000 Reichsmark annually from Flick.

Having recruited the leading Nazi chemist, W.R. Grace Company, having started with Davison Chemical Corporation in 1954 for eleven years, absorbed 23 chemical plants, and then moved to Europe to become the leading grocery producer. Having founded General Foods of Europe, now a full-fledged chemical giant bought up food companies in Europe, becoming the largest distributor, in particular, absorbing the production of pasta in Italy.

At the end of the twentieth century, W.R. Grace Company was convicted with a claim of half a million dollars for violations of environmental standards in 1994 in Montana, later increased by another five million dollars. In 1999, the Environmental Protection Agency filed a lawsuit because of the threat of ammonia poisoning in the water supply system in Michigan; it was also charged for $1.15 billion for 325,000 suits. In 1998, two million were awarded for the victims for the damage from the release of radioactive waste for several decades, and a year later for the same violations in Illinois, another $200 million. Ambros was undoubtedly a particularly valuable employee of IG Farben.

Himmler, who was in charge of the camp system, admired such "remarkable Aryan qualities" of the Indian epos "Mahabharata," where the appearance of a mushroom cloud, comparable to the opening of a giant umbrella, is pictured in the book "Dronaparva." It was a weapon that turned into ashes a whole race of Vrichni and Andhak, whose burned bodies could not be identified, and the survivors fell ill with radiation sickness—they had vomiting, hair and nails fell out. Looking at a similar dreadful cloud of his own making, father of American atomic bomb Robert Oppenheimer also quoted ancient Indian epic: "*Now I become Death, the destroyer of worlds.*" Important to know, however, that the authentic quote from the Bhagavad Gita has the opposite meaning stating precisely: "*I am Death that carries off all things, and I am the source of things to come. Of feminine nouns, I am Fame and Prosperity; Speech, Memory, and Intelligence; Constancy and patient Forgiveness*" (Bhagavad-Gita text 34, chapter 10).

In the ranks of the Third Reich, there was another lover of Eastern antiquity, it was Professor Karl Haushofer, who would visit Hitler and Hess during the imprisonment. At the beginning of the century, he will rise to the rank of general, being an adviser to the Japanese command, he will join the ultra-nationalist organization Black Dragon, which controls Chinese opium supplies and is more known in the modern world as the Yakuza.

It will open like temples of Buddhist monasteries, and participation in the activities of the Order "Green Dragon" will become available, as well as participation in the meetings of the closed British club "Coefficients," which also included Herbert Wells, the author of the term "atomic bomb" used by him in the novel The Liberated World. So who was behind the idea of creating the most deadly weapons in the world, if its development was conducted in different countries, but had a married name: in Germany Projekt U, in the USA Project Y and Japan Project N? The "father of the atomic bomb" for the US Julius S. Oppenheimer was an immigrant from Germany, and he, began work at the Cavendish Laboratory under the leadership of E. Rutherford, where he met with Niels Bohrn and Max Born, then continued his studies at his historical homeland at Goettingen University, where he met W. Heisenberg and E. Schrödinger.

Later, the creator of the American atomic bomb recalled "*the poverty prevailing among the Germans in the mood of extreme humiliation, bitterness, hopelessness, protest and anger that subsequently caused a terrible catastrophe* ...," but the beginning of the last century was the century of German chemistry, and the international scientific language was German. Born in Frankfurt am Main, Otto Hahn—another future Nobel laureate who collaborated with Ernesto Rutherford, returned to Germany, where he became an employee of Emil Fisher, who allocated him his laboratory. Since 1912, he will head a department of radioactive substances at the

Kaiser Wilhelm Chemical Institute, then headed by the "father of chemical weapons" Fritz Haber and will enter the chemical department of the "Haber bureau," directed from 1930 to 1937 by Max Planck and funded by the Rockefellers through the fund headed by Albert Fagler.

It is essential that Albert Fegler would be the founder and in 1939 would replace as chairman of the Supervisory Board created by Dillon, Read and Co., Vereinigte Stahlwerke A.G., in the future Flick Group, which absorbed W.R. Grace Company. Also, Fegler, as president of the Kaiser Wilhelm Society, participated in a crucial meeting in June 1942, under the direction of the four-year plan coordinator, Albert Speer, where Werner Heisenberg spoke about the military application of the atomic bomb and its design.

The Kaiser Wilhelm Society for the Advancement of Science included twenty-one institutes, one of which Fischer studied proteins, teaching the craft of the future Nobel laureate Otto Warburg, whose father Emile Warburg proved the law of photoelectric effect, the Nobel Prize for which was given to Einstein. In 1933 he was the director of the Kaiser Wilhelm Physical Institute in Berlin. In 1888, Fischer married the daughter of a professor of anatomy Agnes Gerlach.

Professor of Frankfurt University Walter Gerlach from 1919 to 1920 headed the physics laboratory Bayer-Werke in the Farbenfabriken Elberfeld, was a colleague of Otto Ambros. Warborgs, like the Oppenheimer, well-known Jewish banking dynasties hailing from Germany, both have to do with environmental research, and if some were closely related to the atomic bomb, the latter are financed by the Third Reich and IG Farben, whose jurisdiction was Auschwitz III. Nobel laureate, Walter Gerlach would become one of the leaders of the German atomic project Uranverein ("Uranium Club").

To extract atomic energy, it was necessary to apply energy to accelerate the alpha particles to overcome the energy of bonds, ensuring the integrity of the nucleus. A step in this direction was the Ernest Orlando Lawrence's cyclotron, in which, the particle was forced to accelerate in proportion to the diameter of the cyclotron. In early 1939 Ernest planned to create an installation, the dimensions of which made it possible to disperse the particles enough to break ties in the atom.

Lisa Meitner, together with Fritz Strassmann, irradiated the slowest neutron flux with the heaviest of the natural elements, uranium, having obtained a result that "contradicts all previous concepts of nuclear physics," as the scientists themselves wrote in the Di Naturvissenshaften on January 6, 1939. Instead of gaining mass, absorbing neutrons, uranium-235 breaks up into elements that are in the middle of the periodic table, which was impossible so far, because the neutron did not possess the energy sufficient for splitting the atom.

Despite participating in the First World War as a nurse, according to racial laws, Lisa Meitner left Germany in 1938 and moved to Stockholm, where Otto Gahn sent her a letter asking "to offer some fantastic explanation." Earlier, for the same reason, her nephew Robert Frisch moved to Denmark with Niels Bohr, while walking through the forest Meitner calculated that the splitting of the uranium-235 atom feeds on the self-energy of the nucleus equal to the mass multiplied by the square of the speed of light or $E = mc^2$. The scientific community in the person of N. Bora in Denmark, J. Frenkel in the USSR, A. Wheeler in the United States gave a detailed explanation of the process of decay.

The idea of the emergence of energy from the vanishing mass opened such exciting prospects that when a former student of Otto Ghana and Lisa Meitner, who now headed the research department of the Auergesellschaft company, Nikolaus Riehl, drew the attention of the German War Materials Department to it. He was reassigned at the Kaiser Wilhelm Physics Institute, and his director was the Nobel laureate, Werner Heisenberg.

Soviet Response

Soviet physicist D. D. Ivanenko also developed the idea of protons and neutrons in the atomic nucleus. It is noteworthy that Heisenberg did much high-altitude cosmic radiation, the discovery of which with the help of balloon measurements was made by the Nobel laureate Victor Hess. Heisenberg's report was the key to the Bohr-organized nuclear congress in Copenhagen.

The calculations of Robert Frisch in 1940 became the basis of the Memorandum "On the Construction" of the Superball "Peierls-Frisch" and the founding of the "Military Application of Uranium Disintegration." The main result of the committee's work was the technical possibility of creating such an atomic bomb.

At one time Vernadsky warned about the possibility of using atomic energy for military purposes, soon in the USSR physicists, Julius Khariton and Yakov Zeldovich determined that a chain reaction is possible if the concentration of isotope 235 is increased and deuterium is used as a moderator. In the first Soviet nuclear laboratory, led by a former consultant of IG Farben. Lange proposed an option for launching a chain reaction and a method for separating uranium isotopes in the required amount.

In 1940, G. N. Flerov and K. A. Petrzhak discovered the spontaneous fission of uranium nuclei predicted by Vernadsky. Scientists of the laboratory of the Ukrainian Physico-technical Institute V. A. Maslov, V. S. Spiegel, and Fritz Lange were filed for a patent application on November 17, 1940: "An aerial bomb or other munitions whose explosion is based on the use of the chain reaction of fission of the nucleus of the uranium-235 isotope at the supercritical mass of the latter, formed by a combination of several subcritical masses, characterized in that in order to format the necessary moment of time the supercritical mass of the isotope uranium-235, the charge of the latter in the bomb is divided into several parts by a series of neutron-impermeable partitions of an explosive, for example, acetyl-silver, which are destroyed by explosion in the required moment " (http://www.geokhi.ru/en/Institute/Vernadsky.aspx). The work was rejected because of the poorly worked out scheme of the uranium charge, but the atomic explosion scheme: the consciousness of the critical mass and the initiation of the chain reaction subsequently became generally accepted.

American Response

Another physicist who left Europe because of racial laws was Leo Szilard. He persuaded Einstein to inform the US President that "an element of uranium is capable shortly… to create potent bombs of a new type." Einstein's letter contained indications that Germany had prohibited the sale of uranium from the mines of Czechoslovakia, and that "the son of German State Secretary von Weizsaecker was seconded to the Institute

Kaiser Wilhelm in Berlin, where now well-known American works with uranium are repeated."

The letter was transmitted to Franklin Roosevelt through Alexander Sachs, who was simultaneously a great friend and unofficial adviser to the president, as well as head of the Department of Economic Research of his National Recovery Administration. Also, he was the vice president of Lehman Brothers, an adviser to the Petroleum Industry War Council, an adviser to the predecessor of the CIA, Office of Strategic Services (OSS). Sachs was an assistant to Louis Dembitz Brandeis, who represented the American Zionist Organization (Zionist) Organization of America. Allen Dallas, the future head of the CIA, was an employee of the law firm Sullivan & Cromwell, serving both Lehman Brothers and Goldman Sachs & Co.

After receiving the letter, in October 1939, Roosevelt issued a decision to establish his "Uranium Committee" (S-1 Uranium Committee), which included himself, and Sachs. A year later, with the assistance of the Joint Chiefs of Staff and the Joint New Weapons Committee (Rear Admiral V. R. Parnell and Brigadier General R. J. Moses), the US president will give an indication of the transformation of the committee into a subcommittee of the National Defense Research Committee (NDRC), which will be headed by the president of the Carnegie Institute, Vannevar Bush, who since 1936 worked on creating the prototype of a modern computer—Rockefeller Differential Analyzer (RDA) for a grant from The Rock Foundation feller.

Bush was also a member of the board of directors of the manufacturer of the uranium rods of Metals and Controls Corporation, now a member of the Texas Instruments chip leader, German pharmacological concern Merck & Co., who later led the decoding of the DNA structure. In 1941, Bush wrote a proposal to Churchill, after which the US and the UK will coordinate their research programs.

June 28, 1941, NDRC ceased to exist, having transformed into the Office of Scientific Research, a year later Vannevar Bush presented to Roosevelt a plan for creating an atomic bomb. In 1943, a student of Niels Bohr, a leading Japanese expert in nuclear physics, Yoshio Nishina, will lead Project N. Under his leadership, the Japanese "uranium committee" recruited Professors Ryokichi Sagane from the University of California, Bunsaku Arakatsu and Masashi Kikuti. Thus began the nuclear arms race.

It began with hunting for specialists, equipment, and consumables for experiments. Specialists of the British Secret Intelligence Service (SIS; MI6) took Niels Bohr to Sweden, and from there in the bomb compartment of the British fighter to the UK. Bohr would later join the US "uranium project" under the fictitious name Baker.

German Counter-Response

An illustrative example of the "uranium race" is the fact of the appearance in Kharkiv Institute of Physics and Technology, Friedrich Houtermans, who proposed a boiler scheme in which the conversion of uranium-235 passes through the stage of neptunium in an element of plutonium. In 1935 he started working at the Ukrainian Physical and Technical Institute with Fritz Lange. In May 1940, Houtermans was deported to Germany, where he began work in the von Ardenne laboratory, where in August 1941 he produced the most complete of the German atomic bomb calculations based on trans-uranium elements.

Based on his calculations, Carl Weizsacker will conclude that it is possible to use "element-94" (plutonium) as an explosive. As early as April 1939, Professor Paul Harteck and his assistant Dr. William Grot informed the Office of the Armed Forces of Germany that the latest achievements in the field of physics paved the way for the production of "an explosive that, in its destructive power, would exceed by many orders of magnitude the explosives of conventional types." The letter went to the Department of Scientific Research and Dr. Kurt Dibner. In September 1939, the Supreme Command of the German Army decided to launch a project to create nuclear weapons.

When in March 1940, Maria Curie's apprentice and son-in-law, Frédéric Joliot-Curie, informed the Minister of Armaments Raul Dautri about the need for heavy water to build atomic weapons, the military intelligence lieutenant Jacques Allier delivered the entire stock of the Norwegian company Norsk Hydro in the amount of 185.5 kg. Norsk Hydro was founded by Hans Olsen, the head of the foreign oil business of Nobels, who was the husband of Mina Nobel. The largest shareholder of the enterprise since the foundation was the Rothschilds, which gives grounds to assume that they were behind Norsk Hydro, primarily since the cartel agreements associated them with the Nobels for a long time.

The Norwegians had robust electrolysis plants for the production of hydrogen, fed by the energy of Norwegian waterfalls, and the creation of heavy water required a large amount of electricity. In May, Germany invaded France, but the assistants of Joliot-Curie, physicists Hans von Holban and Lev Kovarsky, who wrote in March 1939 an article in the journal Nature—"Releasing neutrons in a nuclear uranium explosion," managed to transport heavy water to England (https://www.loot.co.za/index/html/index2395.html).

In September 1939, Adolf Hitler made a speech in Danzig with the words: "*Very soon, there may come a time when we use a weapon that no one can turn against us*" (http://www.hitler.org/speeches/09-13-39.html). The head of "British scientific intelligence" physicist R. V. Jones issued a forecast list that included "… *bacteriological weapons, the latest gases, flamethrowers, airplanes - shells, air torpedoes and unmanned aircraft, long-range missiles, new torpedoes, mines and submarines, deadly rays and magnetic mines* …" (https://www.naval-technology.com/features/featurethe-worlds-deadliest-torpedoes-4286162/).

In 1941, in England, a contract was signed with an annual deadline for the manufacture of a 20-phase model of the isotope separation apparatus by the diffusion method obtained by the firm Metropolitan-Vickers, which took over the consultants Imperial Chemical Industrials. The gas diffusion method, which in the United States will be called Project K-25, is based on the multiple pumping of a gaseous uranium compound through a porous baffle passing an isotope with an atomic weight of 235. The author of the method was the German Baron Manfred von Ardenne, he was not a nuclear physicist, but he was a brilliant experimenter, the author of 600 patents, and a pioneer of television.

In matters of isotope separation Heisenberg relied on heavy water, the lack of a technological line for obtaining which led Kurchatov to use a less efficient graphite retarder. The Americans admitted the presence of German atomic weapons, the historian Wolfgang Schvanitz of Princeton cites a document from the National Archives of the United States, dated January 1945, where the German nuclear program is defined as "strong" and "rapidly developing."

"The systematic bombing of cities in which the laboratories mentioned above and plants were located, as well as the rapid advance of the Red Army and allies, put the German research organizations involved in problem No. 1 in a complicated situation ... Almost all the buildings of laboratories destroyed by bombing and fires, their property was subjected to multiple evacuations. However, the Führer was prevented from snatching victory from the jaws of defeat in "five minutes before midnight" by the rapid advance of Soviet troops and the bombing of "allies." August 23, 1943, ended the Battle of Kursk, the Soviet forces went to the Dnieper, it became apparent that very soon they would be in Europe.

In Germany, since the beginning of 1940, several methods of uranium enrichment have been developed in parallel: the mass spectrometric method, the isotope sluicing method, and the ultracentrifugation method; by the end of the war, 20 tons of "pure uranium preparations" were produced. General V. A. Kravchenko, on November 12, 1945, wrote, "*in Germany, there were about 10 tons of metal uranium with a stock in 1944 of about 500 tons of pure uranium oxide, about 2 tons of heavy water from Norway. Many months after the surrender of Germany in the English zone of the occupation, almost half a million men, still dressed in military uniform, continued to exist in full inviolability; The US military authorities in their zone of occupation began to equip thousands of Polish, Yugoslav and Ukrainian fascists in the guise of "workers' mouths" and all sorts of "security elements*" (Olelinikov 2019).

The Soviet "uranium project" had a shortage of spent uranium. On December 8, 1944, the GKO of the USSR, by resolution No. 7102ss, determined measures to ensure the development of mining and processing of uranium ores the most critical state task. In 1945, a commission of several dozen people was involved in the search for weapons-grade uranium, consisting of Kikoin, Khariton, Flerov, Artsimovich, Nemenov, Golovin, led by Deputy People's Commissar for Internal Affairs Ivan Serov. In addition to the Soviet zone of occupation of Germany, similar work to identify participants and institutions of the German atomic project was deployed on the territory of Austria and Czechoslovakia.

Physicist I. N. Golovin discovered 320 kg of uranium at the Radium Institute of Vienna, which were transported to the USSR together with what was left of the equipment of the Kaiser Wilhelm Institute in Berlin, including the Siemens cyclotron laboratory, and along with Gustav Hertz, the head of the same Siemens laboratory Forschungslaboratorium II. As the head of the Sukhumi object "G" ("Agudzery"), he will be a laureate of the Stalin Prize in 1951, a member of the Academy of Sciences of the USSR and head of the scientific council of the Council of Ministers of the GDR. A thorough investigation helped Khariton and Kikoin to discover 100 tons of uranium oxide, which significantly shortened the term of the atomic bomb project.

Peaceful Atom in Germany

Due to the membership of NATO in 1955, the FRG or West Germany lifted restrictions on the development of specific strategic industries imposed on it after the defeat in the Second World War. Already in 1961, the first nuclear power plant was put into operation here—the "Kal" NPP with the boiling BWR reactor. Its capacity was only 15 MW, but already in 1969 the first commercial nuclear power plant, Obrigheim, began operating with a light-water PWR reactor at an industrial size of 340 MW. These two types of reactors later formed the basis of Germany's nuclear power industry, although there were other varieties in the reactor park—for example, high-temperature gas-cooled, channel and fast sodium.

Nuclear power was a priority in the West German energy policy of the 1960s and 1970s, which only strengthened during the years of the rise in world oil prices and the global energy crisis. The share of electricity generation at nuclear power plants was supposed to be gradually increased to 45%. However, this plan was not implemented, its maximum atomic power in Germany reached by 1990—about 30%.

From the very beginning, the basic principles of the German nuclear power industry were formed. As a rule, nuclear power plant sites are located in the lower reaches and middle reaches of rivers. Accommodation is dispersed, with a link to local consumers in areas that do not have sufficient fuel resources, mainly in the north and south of the country. Therefore, most nuclear power plants have only one-two power units, and the capacity of each station built in the 60–70s does not exceed 1000 MW. This decentralized arrangement helps to equalize the conditions of energy supply and pricing between regions.

Despite the flourishing of nuclear power, there have been some incidents that indicate the ambiguity of its perception by society, which in some cases is even ready to sacrifice irreversible costs incurred. Thus, as soon as it began, the operation of the channel reactor at the Nideraihbakh nuclear power plant was almost completed. A fast sodium reactor at the Kalkar NPP was built but never started. Just a little more than a year worked and was stopped for formal reasons, the nuclear power plant "Mülheim-Kerlich."

However, these alarming bells were lost against the backdrop of the success of the nuclear industry. Germany became the only country outside the Soviet Union that not only built a civil vessel with an atomic power plant—the dry cargo ship Otto Gan, but also for ten years carried out its successful commercial exploitation. In addition to intensively developing nuclear power plants, several dozen research nuclear reactors were operated in the country.

Work was done to develop uranium-plutonium fuel to create a closed nuclear fuel cycle. Kraftwerk Union took the dominant position in the reactor industry (later joined Siemens), which successfully competed even in the foreign market: it was the prime contractor for the construction of ten nuclear power plants in The Netherlands, Switzerland, Argentina, Austria, Brazil, and also started construction Iranian nuclear power plant in Bushehr.

The last nuclear power plant that joined the FRG was in April 1989 the power unit No. 2 of the Nekarvestheim atomic power plant. However, then the political situation changed dramatically.

Nuclear Power Plants (NPP) of East Germany

Nuclear power engineering in the GDR or East Germany has also thrived. The first Rheinsberg nuclear power plant with a 70 MW water-cooled reactor appeared in 1966 with the participation of Soviet organizations, for which it also became the first foreign nuclear power plant. This nuclear power plant has successfully worked for almost a quarter of a century.

Next came the NPP "Nord" (in the West German interpretation—"Greifswald"). It was designed for eight power units with VVER-440 reactors in the years 1973–1979. The first four power units were put into operation; their share in the electricity production of the GDR was 10%. The construction of the next four power units began. Also, the development of two power units of the Stendal NPP with VVER-1000 reactors was started.

The nuclear power industry of the GDR passed away with the state as a result of German unification. The West Germans considered that the technology and technology of the East German nuclear power plants did not comply with Western safety standards, after which all the existing power units were stopped ahead of schedule and construction of new ones was discontinued.

Such a decision looks strange against the background of other examples: VVER-440 reactors are very successfully used at Loviisa NPP in Finland, at Kola NPP in Russia, and at Rivne NPP of Ukraine their estimated service life is extended at once for 20 years (by the way, Rovno NPP is the first in the USSR was tested by the IAEA, highly appreciated its safety). What was more in him: the real concern for security, the political background, or, perhaps, the desire to eliminate competitors?

The East German economy suffered a crushing blow: the four power units of the Greifswald NPP failed to work, the fifth had barely time to enter service in November 1989 (it had only been working for three weeks!), The sixth one was wholly completed, but not commissioned, the construction of the seventh and the eighth was stopped at the high readiness stage.

The situation with NPP "Stendal" was not better: the first unit was built by 85%, the second—by 15%. A proposal was proposed to replace Soviet WWER-1000 reactors with Siemens or Kraftwerk Union on these units, but the government decided unequivocally: the station should not be completed.

A robust uranium mining industry existed in East Germany: large ore deposits were in Saxony and Thuringia, and, of course, were under Soviet control. The joint Soviet-German joint-stock company "Vismut" took the third place in the world in uranium mining and worked with such intensity that after the German unification continued production was unprofitable.

The territory of West Germany was deprived of deposits of natural uranium ores, so its fuel cycle, was only partially formed. The source of uranium ore for Germany was Canada, Australia, and Niger. Uranium hexafluoride was produced in Canada, France, and Great Britain. Enrichment for uranium-235 over time was mastered in Germany. Germany produced fuel fabrication (uranium dioxide pellets, fuel rods) themselves and even had excess production capacity.

Being dependent on foreign uranium ore, the Germans actively worked on closing the nuclear fuel cycle. It was supposed at the factories of France and Great Britain to extract plutonium from spent nuclear fuel from German nuclear power plants and use it for the production of uranium-plutonium fuel. The Siemens Concern even started such production.

The Fate of Fast Neutrons

An essential role in the closed nuclear-fuel cycle was assigned to fast neutron reactors. The first and only fastest sodium reactor in Germany, the SNR-300, was built in 1985 at the Kalkar NPP. However, it was waiting for a sad fate.

The power unit was built 13 years, of which for four years of work stopped due to public protests. Its cost amounted to 7 billion DM (equivalent to 3.5 billion Euros), but he did not work a single day. Just a few months later, the Chernobyl accident occurred in the USSR, and on the wave of social hysteria, a new nuclear power plant was sentenced to liquidation (which cost another 75 million Euros).

The further fate of the Kalkar NPP can be considered a mockery of luck: in its buildings and facilities is located a large amusement park "Wunderland Kalkar," containing more than 40 attractions. In the former cooling tower of an atomic power unit, for example, a large carousel rotates. Every year 600 thousand people visit the park.

Heated by the "green" anti-nuclear sentiments of the public generated protest actions in 1975, for example, the demonstrators captured and held the site for the construction of the Ville nuclear power plant for nine months, which led to the abandonment of its development. Victims of the protests were NPP "Kalkar" and worked only 14 months nuclear power plant "Mülheim-Kerlich," the closure of the nuclear fuel cycle. Chernobyl strengthened these sentiments. Moreover, in the late 1990s, power in Germany came the "green," who actively used for their political purposes, the theme of the fight against nuclear energy. Thus began the decline of the German atomic age.

In 2000, the government and energy concluded an "atomic consensus," based on which the "Law on the Regulated Discontinuation of the Use of Nuclear Energy for Industrial Electricity Production" was adopted in 2002. Its essence boiled down to the following: each NPP was determined the amount of electric power allowed for production and the service life (32 years), for the exhaustion of which it must be closed. The last power unit was supposed to complete its work in 2021.

In 2010, 17 power units remained in operation, which produced 23–28% of all electricity. Given the real situation and difficulties with the introduction of replacement capacities, the government decided to extend the operating time of these nuclear power plants for a period of 8–14 years, that is, the complete elimination of nuclear energy was postponed to 2035. This government decision caused another political bustle, public protests, suits in the federal constitutional court. And then there was an accident in Fukushima, which seems to have dealt a more significant blow to Germany than Japan.

In March 2011 Germany for three months stopped its nuclear power plants built before 1980 and exposed all stations to verification by the Federal Commission for Operational Safety of Reactors. No critical risks have been identified, but the government listens not to the conclusions of experts, but to the opinion of the so-called "ethical" commission that shares the views of the "green."

As a result, out of the 17 operating power units, 8 cease to work definitively, and the remaining 9 should be gradually closed before 2022: NPP Grafenrainfeld in 2015, unit No. 2 of the Gundremmingen NPP in 2017, Philippsburg in 2019, Gronde, Brockdorf and Block No. 3 of the Gundremmingen NPP in 2021, Izar, Emsland and Neckarwestheim in 2022.

All stopped nuclear power plants will be gradually dismantled: this technology has already been developed at the Grosveltsheim, Kal and Niderai-hbah NPPs, which were finally dismantled, and the sites are brought back to their original state. Thus, the nuclear power industry in Germany will cease to exist.

The decision of the authorities caused several lawsuits amounting to more than 18 billion euros from German and Swedish concerns—owners of pre-terminated German nuclear power plants. As compensation for lost revenues, it was proposed to release nuclear companies from payment to the Federal Energy and Climate Fund.

However, the inevitability of the course was confirmed by Chancellor Angela Merkel, saying: "We can become the first large industrial nation to move towards the effective use of renewable energy."

German Wind Race

Today, Germany is the world leader in wind power: 23 thousand wind turbines are in operation here (that is one-third of their number in the world) with a total capacity of 31.3 GW. In the first half of 2012, Germany provided 25% of its energy consumption from renewable energy sources (currently operating nuclear power plants generate 16% of electricity). However, several significant problems are associated with wind power.

Wind stations operate mainly in the north—for example, large wind parks in the North and Baltic Seas, where wind speed throughout the year is at least 5 m/s. The government plans to build another 40 such wind farms.

However, replacing nuclear power is required in the south of Germany. For this, it is necessary to build 3.5–5 thousand km of new transmission lines. Also, these lines should be on direct current. The electricity cannot be transmitted to such a distance without huge losses. In addition to the need for large-scale costs (€20–50 billion), the government faced yet another problem: against the new transmission lines protest.

The second problem is the instability of wind power generation. For example, representatives of the Czech dispatching company CEPS have already stated about difficulties in observing the operating mode of the power system due to the irregularity of energy flows from German windmills. To balance the fluctuations from wind generation, the government encourages research into the creation of smart grids ("smart grids").

Other problems include the connection of offshore wind parks to main power grids, the limited use of windmills in places where there are usually no strong winds, the need to cover winter energy consumption maxima in conditions of calm. Also, the larger the area of the wind farm, the smaller the specific output per one windmill: installations in the wind flow begin to interfere with each other.

The apologists of renewable energy sources consider this as "the difficulties of the transition period" and are confident of bright prospects, but today Germany has to purchase electricity in France and Austria, and also to increase the capacity of its power stations with natural gas (another 30 such units are planned to be built).

Ecological Problems in Germany

The Ministry of Natural Resources of Germany in 2012 recorded an increase in greenhouse gas emissions by 1.6%. At the same time, the growth in the country's economy amounted to only 0.2%, while in metallurgy and the chemical industry—the primary "producers" of greenhouse gases—even a decline to 3.7% was observed. Thus, the increase in carbon dioxide emissions by 18–34 million tons cannot be explained by anything other than the substitution of nuclear power plants by gas power plants. While the operation of all 17 power units of the nuclear power plant made it possible to reduce the release of carbon dioxide into the atmosphere by 150 million tons—according to the deputy of the Bundestag Katarina Reiche, this is comparable to the annual emissions of all cars in Germany.

Several problems related to increasing wind energy, listed in its statement the Land Union for the Conservation of Nature of Baden-Württemberg, equating the harm from windmills to the damage from the extraction of brown coal by the open method. The efficiency of wind energy is extremely low: up to 50% of wind energy is lost on friction, the necessity of accumulating this energy for its uniform transmission also leads to losses, windmills of 200 m height disfigure landscapes and seriously threaten the life of migratory birds (in a year wind turbines in Germany kill up to 220 thousand birds and many bats).

The Uneconomical Economy of Germany

According to economists' forecasts, Germany's renunciation of nuclear energy will cost between €32 billion and €1.7 trillion in the period until 2030 (the upper limit of the estimates is equivalent to more than 65% of the country's annual GDP).

By 2015, according to Deutsche Bank estimates, the average wholesale cost of electricity will increase to €72 per 1 MW • hour (now, after reducing the share of nuclear power plant output from 28 to 16%, this cost already amounted to €60 per 1 MW • hour). According to other estimates, the cost of electricity can increase by 1.5 times, which will increase the value of goods of German manufacturers by 10–20% and reduce their competitiveness in the foreign market.

Household allocations for the development of renewable energy sources in Germany this year increased by 47%, on average from €0.036 to €0.053 per kilowatt hour of electricity consumed: thus, a family of four will pay €250 per year development of wind power. Meanwhile, according to the Consumer Association of North Rhine-Westphalia, in this land alone, 120,000 families have shut down electricity for a year because they could not pay growing bills.

Investments in alternative energy are ineffective. According to Der Spiegel magazine, the €130 billion invested resulted in a 3% increase in electricity production. That is, to completely replace the atomic 28%, will have to spend about €1083 billion. Even if we consider that the initial investment is the largest, the amount will remain impressive. No wonder the Minister of Economy and Technology Philippe Wrestler called investing in alternative energy "a bottomless pit."

Understanding the problems of wind energy, Germany is trying to use all other possible ways of energy supply of the country. The program "SchwarmStrom" for the implementation of 100 thousand gas mini-power stations with a total capacity of 2 GW in the basement of residential buildings was formed. Work is continuing improving wind power plants to make them lighter and less noisy (the problem of sound vibrations also causes discontent among the population).

The authorities encourage residents to save electricity: a system of energy savings checks for households is in place, and those who have reduced household energy consumption receive state preferences.

Nevertheless, while the fact remains unchallenged, in the conditions of abandonment of nuclear power plants, renewable energy sources can provide reliable power supply to consumers only in combination with stations on organic fuel. Here, too, a dilemma arises: Germany plans to reduce the extraction of brown coal and electricity production on its basis, are environmentally unfavorable, and the expansion of gas generation leads to increased dependence on foreign suppliers (for example, Russia).

Thus, Germany's complete refusal from nuclear generation cannot be considered a permanent and effective solution. Perhaps the country will eventually manage to reorganize its "energy basket radically," but the price of this—both economic, social, and even environmental—will be extremely high.

Reference

Olelinikov, P. V. (2019). https://www.nonproliferation.org/wp-content/uploads/npr/72pavel.pdf.

10

The United States as Nuclear Power

The USA is the world's most important largest producer of nuclear power. It accounts no less than 30% of worldwide nuclear generation of electricity. The country's nuclear reactors produced about 20% of total electrical output. Capital expenditure on existing nuclear plants went up in 2012 due to post-Fukushima upgrades, but it went down 26% in 2015 when capital investment in operating plants was $6.25 billion (UNAEA Report 2019). The United States has now 98 operating nuclear power reactors in 30 states, operated by 30 different power companies. Average nuclear generation costs were reduced from $40/MWh in 2012 to $34/MWh in 2017.

US Nuclear Weapons

In 2011, the US armed forces consisted of 450 ground-based intercontinental ballistic missiles (ICBMs), 14 nuclear submarines with 240 ballistic missiles, and about 200 strategic bombers. After the proposed implementation of the START III Treaty, the number of planes would decrease to 60, and the total number of nuclear warheads would be reduced by

D. Basu and V. W. Miroshnik, *The Political Economy of Nuclear Energy*,
https://doi.org/10.1007/978-3-030-27029-2_10

more than three times. The preservation of all three components of the nuclear triad, the qualitative improvement of warheads, and the creation of high-precision strategic carriers in non-nuclear equipment are assumed.

In future, any reductions, the new American nuclear doctrine says, "should strengthen the containment of potential regional opponents, strategic stability in relations with Russia and China and confirm the American guarantees to allies and partners." A new US nuclear doctrine published in April 2010 declares that "the main purpose of the US nuclear weapons is to defend a nuclear attack on the US, its allies and partners. This mission will remain as long as nuclear weapons exist." The United States "will consider the use of nuclear weapons only in emergency situations to protect the most important interests of the United States, its allies and partners."

The United States is "not ready today to endorse a universal policy that recognizes that deterring a nuclear attack is the only function of a nuclear weapon." In this case, it is not disclosed what is meant by the aforementioned unforeseen circumstances. This should be regarded as a serious uncertainty of the US nuclear policy, which cannot but influence the defense policies of other leading states of the world.

US Strategic Offensive Forces

To accomplish the tasks assigned to nuclear forces, the United States has strategic offensive forces (SNS) and non-strategic nuclear weapons (NNW). According to the data of the US State Department, published on May 3, 2010, the nuclear arsenal of the United States as of September 30, 2009, totaled 5113 nuclear warheads. Also, several thousand obsolete nuclear warheads withdrawn from stockpiles were waiting for dismantling or destruction.

The US SNA is a nuclear triad that includes ground, sea, and aviation components. Each component of the triad has its advantages; therefore, the new US nuclear doctrine acknowledges that "*preserving all three components of the triad in the best way will ensure strategic stability with acceptable financial costs with the technical state and vulnerability of existing forces*" (https://fas.org/issues/nuclear-weapons/nuclear-posture-review/).

Aviation Component

The aviation component of the US SNA consists of strategic, or heavy, bombers capable of solving nuclear problems. Their advantage over the ICBMs and ballistic missiles of submarines (SLBMs), according to the new US nuclear doctrine, is that they can be demonstratively deployed in the regions to prevent potential adversaries in crises of strengthening nuclear deterrence and confirming to the allies and partners American commitments to ensure their security. All strategic bombers have the status of "dual purpose": They can strike using both nuclear and conventional weapons.

According to expert estimates, there were about 200 deployed bombers in the aviation component of the US SNA, for which the same number of nuclear warheads was counted (according to the START III rules, now defunct, each deployed strategic bomber is conditionally counted as one warhead, since in everyday activities they all do not have nuclear weapons on board).

Ground Component

The ground component of the US SNA consists of strategic missile systems equipped with ICBMs. The ICBM forces have significant advantages over other elements of the SNA due to the highly secure monitoring and control system, calculated in several minutes of the alert and relatively low cost of combat and operational training. They can be effectively used in proactive and retaliatory strikes to defeat stationary objects, including highly protected ones.

According to expert estimates, at the time of the entry into force of the START III Treaty on February 5, 2011, the ground component of the US SNA consisted of about 450 deployed ICBMs with approximately 560 warheads.

Naval Component

The naval component of the US SNA is nuclear submarines equipped with ICBMs. Their well-known names are nuclear submarines with ballistic missiles (SSBNs) and SLBMs. SSBNs equipped with SLBMs are the most enduring component of the US SNA.

Prospects for the Development of Strategic Offensive Forces in the United States

The current development program for the US SNA does not provide for the construction of new ICBMs, SSBNs, and strategic bombers in the foreseeable period. At the same time, by reducing the total reserve of strategic nuclear weapons during the implementation of the proposed but defunct START III Treaty, "*the United States will retain the ability to 'load' a certain number of nuclear warheads as a technical safety against any future problems with delivery systems and warheads, as well as in the event of a significant deterioration in the security environment*" (http://www.armscontrol.org).

In the area of developing nuclear weapons equipment, the initial efforts in the United States in the coming years will be directed at improving existing nuclear warheads. Launched in 2005 by the Department of Energy under the Reliable Replacement Warhead (RRW) project, the development of a highly reliable nuclear warhead has now been suspended.

As part of the implementation of the non-nuclear prompt global strike strategy, the United States continues to develop technologies for guided warheads and warheads in non-nuclear equipment for ICBMs and SLBMs. These works are conducted under the leadership of the Office of the Minister of Defense (Advanced Research Administration), which eliminates duplication of research conducted by the types of armed forces, more efficiently expend money and ultimately accelerate the creation of high-precision combat equipment for strategic ballistic missiles. Since 2009, several demonstration launches of prototypes of intercontinental delivery systems are being carried out, but no significant achievements have yet been achieved. According to expert estimates, the creation and deployment of high-precision ICBMs and SLBMs with non-nuclear equipment can hardly be expected before 2020.

Non-Strategic Nuclear Weapons in the United States

Since 1992, the United States significantly reduced its arsenal of non-nuclear weapons (NNW). As emphasized in the new US nuclear doctrine, today the United States retains "only a limited number of advanced-based nuclear weapons in Europe, as well as a small number in US warehouses in readiness for global deployment in support of enhanced deterrence in the interests of allies and partners."

US Nuclear Reductions in the Future

The new US nuclear doctrine states that the President of the United States instructed to review possible future cuts in American strategic nuclear weapons below the levels established by the proposed but now defunct START III Treaty. It is emphasized that several factors will influence the scale and pace of subsequent reductions in US nuclear arsenals.

First, "*Any future cuts should strengthen the deterrence of potential regional opponents, strategic stability in relations with Russia and China and confirm the US security guarantees to allies and partners*" (http://www.armscontrol.org).

Second, "*The implementation of the program: Maintaining readiness of the nuclear arsenal (Stockpile Stewardship Program, SSP) and recommended by US Congress, financing of the nuclear infrastructure will allow the US to abandon the practice of having a stockpile of un-deployed nuclear warheads in the reserve in case of technical or geopolitical surprises and significantly reduce due to this, the nuclear arsenal*" (http://www.armscontrol.org).

Third, "*Russia's nuclear forces will remain a significant factor in determining how much and how quickly the US is ready to further reduce its nuclear forces*" (http://www.armscontrol.org).

Assessing these intentions of Washington, it should be noted that they practically do not take into account the concerns of Moscow, caused by the deployment of the American global missile defense system, which in the future can weaken the deterrence potential of Russia's strategic nuclear forces (SNF); the tremendous superiority of the United States in conventional armed forces, which may increase even further with the adoption of the developed American systems of precision-guided long-range weapons; as well as the reluctance of the United States to support the draft treaty banning the deployment of any weapons in space introduced by Russia and China to the Conference on Disarmament in Geneva in 2008. Without finding mutually acceptable solutions to these problems, Washington is unlikely to be able to persuade Moscow to renegotiate further reduction of nuclear arsenals.

In 2011, the US armed forces consisted of 450 ground-based ICBMs, 14 nuclear submarines with 240 ballistic missiles, and about 200 strategic bombers. If the START III Treaty, now defunct, would be implemented, the number of planes will be reduced to 60, and the total number of nuclear warheads will be reduced by more than three times. Preservations of all three components of the nuclear triad, qualitative perfection of warheads, as well as the creation of high-precision strategic vehicles in non-nuclear equipment, are envisaged.

Shortly, any cuts, the new American nuclear doctrine says, "should strengthen prevention of potential regional opponents, strategic stability in relations with Russia and China and confirm American guarantees to allies and partners" (http://www.armscontrol.org/factsheets/pniglance).

In the new US nuclear doctrine published in April 2010, it is declared that "the primary purpose of US nuclear weapons is to defend nuclear attacks on the United States, its allies and partners". This destination will remain so as long as there are nuclear weapons. According to the US government, it will consider the possibility of using nuclear weapons only in emergency circumstances to protect the vital interests of the United States, its allies and partners.

The United States "is not ready today to endorse a universal policy that recognizes that deterring a nuclear attack is the only function of nuclear weapons" (https://fas.org/issues/nuclear-weapons/nuclear-posture-review/). However, it is not disclosed what is meant by the aforementioned unforeseen circumstances. This should be regarded as a serious uncertainty of US nuclear policy, which cannot but affect the defense policy of other leading states of the world.

To fulfill the tasks assigned to the nuclear forces, the United States has SNS and NNW. According to the US State Department, published on May 3, 2010, the United States nuclear arsenal as of September 30, 2009, had 5113 nuclear warheads. Also, several thousand obsolete nuclear warheads, withdrawn from the reserves, were expected to be dismantled or destroyed.

The Decline of US Strategic Nuclear Forces

If we assess the state of the US SNF for the past 20 years, we can conclude that they have been fully and deeply degraded, which since 2005 has become irreversible and uncontrollable. In 1990, when it became clear that the Soviet Union was living out the last days, the United States accepted the program of developing SNF until the year 2000. Despite the descent from the historical arena of its main rival in the person of the USSR, it was decided to keep the existing strategic nuclear potential in quantitative terms and significantly strengthen it in qualitative terms. That was a well-grounded decision—the rising star of China was already looming on the horizon.

By 2000, the Americans planned to increase the grouping of ICBMs a little—from 1000 to 1050. However, the main thing was not quantitative, but in its qualitative composition.

By 2000, the United States planned to have 480 submarine-launched ballistic missiles (SLBMs) deployed on 20 nuclear submarines (SSBNs). At the moment, only 14 SSBNs with 336 SLBMs make up the sea component of the US SNF (losses from the planned figure were 30%). It is undoubtedly the most battle-worthy component of the US nuclear triad, although the Trident II D-5 SLBM's operational lifetime has already exceeded a quarter of a century.

With the US Navy's standard for alert duty (60% of the time), in the next ten years, US SSBNs will be locked in bases due to their inability to go to sea. The combat readiness of the SLBM Trident II at this time also causes great doubts, because, being solid fuel.

Of course, conclusions can be made ambiguous. However, here are the options. Shortly, the degradation of the US SNF will continue, and within 10 to 15 years, America will withdraw from the world arena of political heavyweights, turning into a regional power, sluggishly defended by tactical nuclear weapons from its neighbors.

Officially, the United States stopped producing nuclear munitions in 1993, and weapons-grade plutonium has not been produced since the late 1980s

An assessment of the state and prospects of the development of US nuclear forces allows us to assert that Washington is implementing a large-scale and very costly program to maintain their combat readiness, construction, and development. This program also provides for qualitative improvement of the US nuclear arsenal with its quantitative reduction. To this end, the modernization of the US nuclear weapons complex (NQO) is planned, which, as noted in the Nuclear Review, has been forgotten.

To correct the current situation in the US Nuclear forces, it is envisaged to implement the following recommendations, given in the Nuclear Review (https://fas.org/issues/nuclear-weapons/nuclear-posture-review/).

- it is necessary to strengthen the scientific, technical and engineering bases of nuclear weapons, which play an important role in maintaining the preparedness of the nuclear arsenal and enhancing the integrated operational security of nuclear weapons and its certification without nuclear testing;
- for long-term maintenance of a nuclear arsenal reliably protected from unauthorized access and use, it is necessary to increase the financing of personnel of nuclear weapons, as well as scientific and production funds;
- to finance the construction project at the Los Alamos National Laboratory of the chemical and metallurgical research complex for the replacement in 2021 of an existing fifty-year facility;
- to create a new uranium processing plant at the Y-12 plant in Oak Ridge, Tennessee, and to ensure its commissioning in 2021.

All of the above makes it possible to conclude that, despite the rhetoric from Washington about the desire for a nuclear-free world, in fact, the US military and political leadership is not going to abandon nuclear weapons.

The California Institute for International Studies in Monterey (The Monterey Institute of International Studies) and the James Martin Center for Nonproliferation Studies conducted a study on the allocation of costs for the maintenance of US SNFin a state of combat readiness in for the next 30 years. During this period, the Americans plan to spend about one trillion dollars for these purposes, which are supposed to be spent on the purchase of new nuclear weapons carriers, improvement of the armament nuclear warheads, and the head of the ICBMs.

Before the start of the sequestering of the defense budget, the Obama administration planned to replace the existing systems at a faster pace. However, analysts note that the new procurement schedule carries significant risks and is likely to lead to even higher costs, lesser combat capabilities, and a slower deployment of the replaced components of nuclear forces.

The projected cost of technical support for weapons in the system, taking into account the programs for extending the life cycle of nuclear warheads, as well as the necessary replacements in each of the components of the US nuclear triad can range from $872 billion to $1082 trillion. The estimated annual cost for maintenance of the SNF components is $8–9 billion. At the same time, according to the data of the budget and finance department of the US Congress, they can spend annually 12 billion dollars to support the SNF. Twelve billion dollars can be spent on next-generation systems that can replace the components of the nuclear triad currently in service. Simultaneously, about $8 billion in the budgets of the Navy and the US Air Force can be required for the maintenance of new SNF.

Following the new Treaty on Strategic Offensive Arms (START), the United States plans to maintain the combat capability of all 14 Ohio-type SSBNs with 240 SLBM deployed on them, with simultaneous conversion or complete removal of four launching pits on each boat.

The annual cost of supporting the naval component of SNF for the period of implementation of the future program of the Future Years Defense Program (FYDP) ranges from $2.9 billion to $3 billion or $14.6

billion for the entire period under review. These costs cover technical support for SSBNs and SLBMs, without taking into account personnel costs, long-term costs for removing the boat from weapons and decommissioning the reactor, pensions and medical expenses for outgoing members of the armed forces.

By the plans of the Ministry of Defense of the Ohio SSB, it is planned to withdraw from the fleet from 2027 to 2042. The removal of the SSBN from the armament will occur with the pace of one boat during the year. The US Navy is going to replace the existing SSBN with 12 SSBN (X) boats. Current funding for the SSBN (X) program is aimed at developing technologies, including a unified missile compartment and a full electromotive system.

The total cost of replacing SSBNs on SSBN (X) is estimated at $77–102 billion, with the price of one boat at $7.2 billion. The Navy plans annual operating and maintenance costs for each SSBN (X) at $124 million, or nearly $1.5 billion for 12 boats. At the same time, the Navy hopes to reduce the cost of both the ship itself and the costs of its operation and support. It is planned to allocate $6 billion for R&D, as well as $1.6 billion for advance purchases.

The annual costs for this fleet can be $3.1–3.5 billion during 2014–2018 or $16.5 billion. At the same time, according to the BBC's budget, this program can require $10 billion over the next five years from 2018.

According to the 30-year plan for the annual financing of the Air Force and acquisitions of aviation equipment published in 2012, $55 billion is allocated for procurement, and this type of armed forces plans to purchase 80–100 new bombers. These estimates do not take into account R & D, while independent analysts estimate the cost of this item from 20 to 45 billion dollars. Taking into account the data outlined in the 2006 US Congressional Budget Office study, the total cost of the long-range subsonic bomber program can be 92 billion dollars, of which $61 billion can be needed for procurement and $31 billion for research and development.

A specific part of the costs of ensuring SNF in the next 30 years will have to work to maintain the service life of nuclear warheads conducted by the administrative bodies for ensuring nuclear safety. These works are carried out under the LEP (Life Extension Program) and will require expenditures of $70–80 billion.

In general, as noted in the report, the United States will spend about one trillion dollars from 2013 to 2042 to support SNF and purchase a new generation of nuclear-armed weapons.

Despite the plans to reduce the budget deficit by $1.2 trillion in the next 10 years from 2018 and the ongoing negotiations of the so-called super-commission on reductions of another $ 1.5 trillion, members of Congress are pushing the construction of a complex at the Los Alamos National Laboratory in New Mexico for storage and auxiliary production of plutonium. Critics say that the complex is not so necessary, poorly designed, and dangerous—fault lines penetrate the entire territory of the facility—and its cost has increased from 375 million dollars in 2001 to US$5.5 billion in 2018.

US Nuclear Industry and Reactors

The United States was a pioneer of commercial nuclear power development after the Soviet Union, with Westinghouse designed the first fully commercial pressurized water reactor (PWR) of 250 MW capacity, in 1960 and operated up to 1992. Meanwhile the Argonne National Laboratory has developed the BWR (boiling water reactor), and the first commercial plant, Dresden 1 (250 MW) designed by General Electric, was started up in 1960. A prototype BWR, Vallecitos, was operated from 1957 to 1963.

By the end of the 1960s, reactor units of more than 1000 MW capacity, for PWR and BWR, were constructed. Although nuclear developments in the United States suffered a major setback after the 1979 Three Mile Island accident, though that actually proved the efficiency of the design principles of reactors, as none was injured or exposed to harmful radiation. By 1990 over 100 commercial power reactors were commissioned.

In the United States, there are zones of both regulated and competitive electricity markets. About half of 99 currently operating nuclear power units were in the regions where liberalization was carried out. In areas of competitive markets whose prices are under pressure due to the slow recovery in demand for electricity, lower gas prices increased the competitiveness of other types of generation due to the development of technology and the provision of preferences to stations on renewable energy sources; nuclear generation generally feels worse.

It is no accident that a number of nuclear power plants in the market liberalization zones are closed by their owners before the expiration of the generation license. This is the only reason why two nuclear power plants (Vermont-Yanki and Kivuni) were closed in the United States; already announced the early closure in the next 2–3 years, two more (Pilgrim and Fitzpatrick), as well as the possibility of such a decision in relation to several more stations (Kuod City, Byron, Clinton, Davis Bess, Jeanne).

However, it should be added that these fossil fuel stations affect even more. So, according to the research company Velocity Suite, nuclear power plants account for only 6% of the capacity, closed in the United States in 2010–2015. During the same period, 41.3 GW of coal generation and 18.4 GW of gas were closed (compare with 5 GW of nuclear). According to the US industry association NEI, due to the tightening of environmental standards until 2020, their fate can be divided up to one-third of the US coal production capacity. Against this background, the situation of nuclear power, which has a low cost of generating electricity and not bearing the cost of emissions, does not look so dramatic. It is no coincidence that representatives of some companies-owners of nuclear power plants (for example, Exelon) said that they are concerned not so much with competition from gas and, notably, coal stations, as excessive privileges of RES.

Although the total volume of various forms of federal subsidizing of nuclear energy cannot be called insignificant (about $1–2 billion per year), compared to the scale of the world's largest nuclear generation (99 power units), state support looks rather modest. Until the present decade, such support has been reduced almost exclusively to federal subsidies to R&D. According to the US research company Management Information Services Inc., in 1950–2010 the state allocated $73.8 billion for this purpose, which amounted to less than 9% of the total energy subsidies (although about half of all energy R&D projects).

In recent years, the arsenal of state support instruments has expanded. In 2005, the Federal Law on Energy Policy was adopted, declaring several additional measures, including those directly related to nuclear generation. These include: tax deductions for electricity generation at planned (now under construction) reactors, state financial guarantees for projects to build new capacities in the industry, some reduction in allocations to decommissioning funds, measures to compensate for a number of risks, additional financial support for R&D. However, most of these measures do not apply to all, but only to individual projects selected according to particular criteria.

In particular, the right to tax deductions for future generation (about $0.02/kWh in the first eight years of operation, which is equivalent to about half of the current average cost of nuclear generation in the United States) can only be obtained by the owners of four power units being built at nuclear power plants "Vogtle" and "Summer." The fact is that the range of special projects is determined by the terms of the introduction of reactors, into which only the named blocks are inscribed. The same with federal state guarantees: for today they are approved only for investors of the second stage of NPP "Vogl" (in the total amount of $8.3 billion). To allocate further among the remaining candidates in the nuclear industry (not only generation) another $12.5 billion has been awarded.

Problems of the Nuclear Industry in the United States

The first experimental fast-neutron reactor was created in the United States. On December 20, 1951, the EBR-I (Experimental Breeder reactor) reactor at the Argonne National Laboratory in Idaho for the first time produced enough electricity to power four incandescent lamps, and the next day produced enough energy to provide free food for the whole building in which it was located. Of course, by modern standards, it was a completely "toy" reactor, but experiments on it continued until 1964.

Power EBR-I was about 200 kW, but due to its inconsistent operation and several accidents (for example, in 1955 it even had a partial melting of the core), it was never connected to the network. The melting of the base was caused by the unreliable operation of the selected coolant of the primary circuit, the eutectic (mixture) of liquid sodium and liquid potassium, which was subsequently abandoned.

Therefore, a fast-neutron reactor may well qualify for the title of the first reactor, which, in addition to heat, also supplied us with electricity. The next generation of experimental reactors was introduced in the United States by the EBR-II reactor, which was commissioned in the Argonne National Laboratory in 1964 and operated until 1994. It was already developed with an eye to the closed nuclear cycle—when the EBR-II was created, it was officially declared full reprocessing of spent fuel, separation of residual uranium and plutonium from it, and reuse of these isotopes on the EBR-II site.

However, the reactor honestly worked for 30 years and produced an average of about 20 MW of capacity, with the planned maximum heat capacity of 62.5 MW. This useful power of the reactor ultimately provided for much of the thermal and electrical energy for the objects surrounding the reactor.

That is, already at the EBR-II project, engineers and designers faced two "successes" of the cycle and a fast-neutron reactor: The task of processing fuel and the responsibility of monitoring the state of the reactor were not so trivial as they seemed at the very beginning.

However, the EBR-II technology has already been fully developed technology of liquid metal coolant, which was already well-known to us pure metallic sodium, which replaced the failed unsuccessful potassium-eutectic eutectic.

Interestingly, on EBR-II, the "passive safety" system has already been thoroughly tested. In April 1986, the EBR-II conducted a series of tests on "passive safety." In the first test, the main pumps of the primary circuit cooling system were disconnected entirely, while the reactor was at full thermal power (62.5 MW). By not including emergency cooling systems, station personnel was able to dump the reactor for about 300 seconds (5 minutes). No damage to nuclear fuel or reactor structures was recorded. This test showed that even with the loss of all electrical energy and the impossibility of stopping.

The world's first commercial fast reactor LMFBR (Liquid Metal Fast Breeder Reactor) was also built in the United States at the Enrico Fermi station in Michigan. The "Fermi-One" power, as this unit was called, amounted to 94 MW of electric power. The construction of the "Fermi-One" started even earlier than the construction of the EBR-II, back in

1956, immediately on the wave of success of the first American breeders and the opening of a station in Obninsk in the USSR. The foundation of the Fermi-One block was laid in 1956. However, the materials were not worked out; there was no definiteness with the stability of the design in the high-energy neutron fluxes. The deadlines were constantly postponed and American nuclear power plant, which gave a current to the grid, gradually went to traditional thermal neutron reactors, the first of which gave a current to the network at the nuclear power plant Shippingport in 1958.

"Fermi-One" also became an incredible long-term construction and was put into operation only in 1963, seven years after the construction began. However, almost immediately on the "Fermi-One" flawed flaws and design shortcomings, which jeopardized the very existence of the reactor. So, on October 5, 1966, pumping of sodium was blocked by a piece of zirconium, which broke away from the fuel assembly and blocked the channel for pumping sodium. As a result of local overheating, six fuel assemblies wholly melted and the entire first circuit of the reactor was contaminated with radiation. This case itself became the basis for the book of John G. Fuller with the title "How we almost lost Detroit." The detached piece of zirconium was removed from the reactor only after a year and a half, as early as April 1968. Two years later, the damaged core was repaired, and the nuclear power plant was ready to resume work in May 1970, but suddenly a fire with a sodium coolant appeared at the station. The nuclear power plants decided to restart in July 1970. However, the station was never put into operation until August 1972, when the operating company Fermi-One was finally denied the license for the long-suffering block.

However, the American breeder program did not end with Fermi-One. The final point on the issue of American breeders was put correctly by the project of the breeder reactor in the Clinch River. It was The Clinch River Breeder Reactor, abbreviated CRBRP and was to become that "soaring bird," which was intended to make the breeder program of the United States to the unattainable heights of the closed nuclear cycle. The goal of this project was the development and construction of a fast neutron sodium reactor, which was to be a demonstration prototype for the next class of similar American reactors. At the same time, the Clinch River reactor was conceived as an essential step on the way to mastering the technology of liquid metal fast reactors for commercial use in the electric power industry.

The location of the Clinch River reactor was to be an area of 6 km^2, administratively part of the city of Oak Ridge in Tennessee. The Clinch River reactor was supposed to have a thermal power of 1000 MW and electric power in the range of 350–380 MW. In this respect, it was to become the "sibling" of the Soviet BN-350 breeder reactor, which had already been built in Kazakhstan's Shevchenko (now Aktau). Fuel for the reactor was to be 198 hexagonal assemblies assembled in the form of a cylinder with two fuel enrichment zones. The internal part of the reactor should consist of 108 meetings containing plutonium, enriched to 18%. They were to be surrounded by an outer zone composed of 90 assemblies with plutonium improved to 24%. Such a configuration should provide the best conditions for heat release. Thus, the swing in the case of CRBRP was generally infernal—it should have been the world's first purely plutonium reactor-breeder.

For the first time, the CRBRP project was introduced in 1970. In 1971, the US Atomic Energy Commission found that the cost of the project would be "about $400 million." The private sector promised to finance most of the project, allocating about $257 million for these needs. In subsequent years, however, against the backdrop of the initial development and refinement of the project price, the cost of designing and building the CRBRP jumped to 700 million. As of 1981, after eleven years of design and development work and mass testing, the CRBRP project has already spent a billion dollars in budget funds, while the total cost of the reactor construction project was estimated at $3–3.2 billion, not counting one billion more, which was necessary for the construction of a plant for the production of special plutonium nuclear fuel. Before deciding to close, the project cost was estimated at 8 billion US dollars.

Another reason was the high cost of construction and operation of the reactor itself, the breeder for the production of electricity. In 1981, it was estimated that the cost of building a fast reactor would be twice the construction of a standard light-water reactor of the same capacity. It was also determined that for a breeder to compete economically with conventional light-water reactors, the price of uranium should be $165 per pound, while in reality, this price was then only $25 per pound. Private generation companies did not want to invest in such a risky technology.

The Clinch River project was resumed after Ronald Reagan came to power in 1981. Despite growing opposition from the Congress, he abolished the ban on his predecessor, and construction was resumed. However, on October 26, 1983, despite the successful course of construction work, the US Senate called for a majority (56–40) to abandon further financing of construction, and the facility was evacuated.

In the 1990s the United States was in a halo of its power, an HEU-LEU agreement had just been concluded that would allow Washington to simultaneously destroy the reserves of Russian highly enriched uranium (HEU) and to import cheaply obtain low-enriched uranium (LEU) from Russia for its needs.

However, Russian enrichment capacity proved to be much more useful than the Americans assumed. Further, market laws began to work against Americans. Instead of destroying Russian enrichment plants, the United States had to close its own.

The construction of the Temelin NPP began in 1985. By 1990, it was decided to suspend its construction in connection with the "insufficient reliability of the Soviet bloc's." However, the shortage of electricity forced the Czech leadership to review the strategy—it was decided to finish building two units using the developments of the American company Westinghouse.

The first reactor was launched in 2000 and the second reactor in 2002. Both were commissioned in 2002 and 2003, and in 2007 problems began. The fact is that from the very first loading of fuel into the reactors, deliveries were made not by the Russian company TVEL, but at that time by British, recent Americans and future Japanese (Westinghouse Electric Company). The Temelin NPP has become an experimental ground for creating an alternative to Russian fuel for VVER-1000 reactors.

After several years of operation, deformation of fuel assemblies (fuel assemblies) was revealed. The fact is that in VVER-1000 reactors, the operating modes differ from the working methods of American ones. The management of the Czech Republic was so scared that their fuel was unloaded to full burn-out, and both reactors were rebooted by Russian fuel.

A little earlier, experiments began in Finland on VVER-440 units. In 1998, Westinghouse first attempted to master the fabrication of fuel cassettes for these reactors. At that time, five fuel assemblies for the experimental operation were delivered to the Loviisa nuclear power plant. They were loaded together with six Russian fuel assemblies. In 2002, all 11 cassettes were unloaded and investigated.

The corrosion resistance of Westinghouse cassettes was significantly lower than the Russian ones. The corrosion layer in American assemblies was 21 μm, while for Russian meetings it was only 1–4 μm. Also, "American TVELs" recorded the release of gaseous fission products, twice the equivalent of Russian fuel assemblies. As a result, the owner of the station, Fortum, decided to cease cooperation with Westinghouse, and so far no experiments have been done on the VVER-440 units

Even before the accident at the Temelin NPP in August 2005, the first block of fuel from Westinghouse was loaded in the third unit of SUNPP instead of traditional fuel from TVEL. This program was agreed upon under Ukrainian president Kuchma when in 1999 Westinghouse co-founded the Center for Designing Active Zones based on the National Science Center "Kharkov Institute of Physics and Technology."

This decision was the beginning of the project "Nuclear Fuel Qualifications for Ukraine" as part of the agreement between Ukraine and the United States in 2000 to diversify nuclear fuel supplies. In 2012, on the seventh year of fuel exploitation, its deformations were noticed. In the eighth year, they became critical, and US fuel assemblies were hardly extracted from the reactor.

By the way, in the Czech Republic, the problems were also revealed in the seventh year of fuel exploitation, which is not accidental. The processes in the reactors are the same. In general, such stability of the emergence of problems shows that they are not random.

After this incident, cooperation with the United States was curtailed, despite the conclusion in 2008 of an agreement to transfer a minimum of three (six) Ukrainian VVER-1000 to US fuel in 2011–2015.

For the American company Westinghouse Electric, hard times came. It was not from a working life that it began its struggle for the Ukrainian and Eastern European nuclear market, trying to shove its fuel assemblies into the Russian VVER-1000 reactors. Only after snatching a piece of

Russian nuclear pie, the company could avoid bankruptcy. The idea to bite off from Russia a portion of its atomic pie originated in the West immediately after 1992. His nuclear power plants, built in the 1980s, are a delicious pie, which also had to be swallowed. Then it seemed to many that Russia would never rise again and it would be necessary to snatch away more tidbits from it.

In the 1990s, it all seemed simple: The United States was in a halo of its power, an HEU-LEU agreement had just been concluded that would allow Washington to simultaneously destroy the reserves of Russian HEU and cheaply obtain LEU for its needs. The fact that this eventually turned into a long-lasting catastrophe became clear much later when nothing could be done about it. Russian enrichment capacity proved to be much more useful than the Americans assumed. That is why the agreement, which was to become deadly for the nuclear industry of Russia, grew for her salvation in the most challenging 1990s. Further, market laws began to work against Americans. Instead of destroying Russian enrichment plants, the United States had to close its own.

However, it was found out later, but so far in the late 1990s, the Americans had reached the home stretch, creating nuclear fuel for Soviet reactors, which was supposed to inflict the final blow to the Russian nuclear industry.

Failure of the Westinghouse Electric Co.

Westinghouse Electric is the leading nuclear energy corporation in the United States, building nuclear power plants globally, mainly in emerging countries. In 2006 Westinghouse Electric was acquired by Toshiba. In 2017 the Westinghouse Electric Co., which was under Toshiba Corp umbrella, filed for bankruptcy saying the "*nuclear renaissance was not realistic when the US 'lost a lot of expertise in the nuclear power plant construction.*" (April 24, 2018, Mainichi Japan).

By 2006 nuclear power was gaining attention as a countermeasure to tackle global warming, particularly in the United States. However, the Fukushima No. 1 nuclear power plant disaster in 2011, demand for new nuclear power plants collapsed worldwide. These problems have afflicted nuclear power plants around the world. 175 out of the 180 nuclear projects worldwide exceeded the initial budget, on average by 117 percent, and took on average 64% more time than projected. An essential source of the technical problems has been the reactor coolant pumps (RCPs) that were supplied by US manufacturer Curtiss-Wright Corporation.

Westinghouse Electric top management thought that problems of the cost could be solved using it's AP1000 reactor design. Westinghouse projected that the AP1000 reactor would require an accelerated construction period of approximately 36 months. However, as Westinghouse admitted to the US Nuclear Regulatory Commission later, the development of the project was incomplete; nevertheless, the company rushed and accepted a half-baked plan. Then the project made by Westinghouse went wrong in China, in Sanmen and Haiyang, as well as in the United States. Thus, Westinghouse might not get any more contracts to build nuclear reactors in China. The construction of both projects in China was abandoned, as well as other twelve Westinghouse reactors were left in the United States between 1977 and 1989 at various stages of development (Global Nuclear Power Database in https://thebulletin.org/global-nuclear-power-database/). General Electric recently has begun to ramp back plant construction, and the French nuclear company Areva, once a leader in the industry, is also in the process of restructuring, as a fuel supplier. However, the future of peaceful atom and nuclear power in the United States is mostly uncertain.

Disruption in the Nuclear Industry in the United States

In 1971, the first leaks in the PWR steam generators caused by the corrosion of alloy 600 were recorded. Then, on the block No. 3 of the San Onofre NPP with the PWR reactor, a nozzle for the instrumentation of the instrumentation on the volume compensator was corroded. In 1987, the sleeves of the same compensator flowed on the block with PWR in Arkansas. And in 1989, the first information appeared on the corrosion of PWSCC on French blocks. Two years later, the first cases of corrosion damage to the nozzles for the entry of CPS drive on the lids of Western reactors under pressure were discovered.

In Germany, after known problems at the Obrigheim nuclear power plant, it was decided to abandon the 600 alloys in favor of the Incolloy 800, which allowed local nuclear scientists to remain outside observers in the ensuing people of replacing the out-of-order equipment of the primary circuit.

In the United States, civilian nuclear power has always received support from the state. During the Cold War, nuclear power plants received fuel from the military in exchange for spent nuclear fuel from which they subsequently extracted plutonium for nuclear munitions. It was then that 104 US reactors were built. Now, so much plutonium is not needed, the massive construction of nuclear power plants in the United States ended after the accident at the nuclear power plant Three Mile Island in 1979. Nevertheless, the nuclear industry continues to receive support from the state, mainly at the expense of mitigating security requirements and by providing government guarantees for loans and credits.

President George W. Bush in 2007 announced the launch of a multi-year advertising campaign for the construction of a "new generation" nuclear power plant. President Obama promised to guarantee the borrowed funds for the expansion of Vogtle NPP for 8.33 billion dollars from the US Department of Energy. The developer of the reactor AP-1000, the American corporation Westinghouse (Westinghouse), advertises its product as "simple, safe, and innovative." The reactor is a further development of the technology of "water-water" power reactors, similar to the Russian WWER and French EPR.

As one of the main advantages of Westinghouse is to simplify the design, reduce the number of components and "modularity" of the structure, as well as the existence of "passive" security systems. All this makes the project similar to the Russian WWER-TOI, which will soon be widely advertised by ROSATOM.

The AP-1000 has two cooling loops, 3400 MW of thermal power and 1100 MW of electrical power. NPP's efficiency is 32.4%. Thus, like its analogs, only one-third of the thermal energy produced by the reactor is used. Two-thirds of the power—more than 2000 MW will be emitted into the atmosphere through the cooling towers.

> The mission of the National Administration for Nuclear Safety, which is a semi-autonomous agency of the US Department of Energy, was established to "*strengthen national security through the military use of nuclear energy.*" (https://nsarchive2.gwu.edu/)

Initially, the activities of the NNSA were focused on the reconstruction of the building of the old chemical-metallurgical research complex, by the 1990s, obsolete and dilapidated. Studies have revealed the presence under the object of tectonic faults, which can cause dangerous earthquakes. As follows from the reports of American scientists, as of June of 2018, the United States had 1950 operational-strategic nuclear warheads, plus about 200 deployed on behalf of the allied countries—Belgium, Turkey, the Netherlands, Italy, and Germany—and 2850 in reserve. Also, approximately 3500 warheads need dismantling. All this is consistent with the figures given by the State Department in 2010. START I requires the United States to reduce the number of deployed warheads to 1550 by 2018; however, President Trump has canceled that Treaty. At the peak of the Cold War—in 1967—the country's nuclear stock was 31,225 warheads. By 1989, America had 22,217. So in this respect, the American arsenal is well provided but lacks new development as Russia has demonstrated already. That could be the possible reason for President Trump to remove the restriction on research and development imposed by these treaties with the Soviet Union and Russia.

11

Russia as a Nuclear Power

"Peaceful Atom" in Russia: Recent Developments

The Russian nuclear industry is one of the most advanced in the world in terms of scientific and technical developments in the design of reactors, nuclear fuel, nuclear power plant (NPP) operational experience, and access to NPP personnel. The industry has accumulated vast experience in solving large-scale tasks such as the creation of the world's first NPP (1954) and the development of fuel to use in it. Russia has the most advanced enrichment technologies in the world and projects in which NPPs use water-cooled power reactors have proven their reliability via thousands of reactor-years of trouble-free operation. The nuclear industry can act as a driving force for the development of other sectors. It provides orders and therefore represents a developmental resource for machine building, metallurgy, buildings and infrastructure, and other industries.

D. Basu and V. W. Miroshnik, *The Political Economy of Nuclear Energy*,
https://doi.org/10.1007/978-3-030-27029-2_11

ROSATOM Is an Absolute Global Leader in the Nuclear Industry

The Russian nuclear industry is a world leader in the peaceful use of atoms thanks to its retraining of the many cadres it still has from the days of the Soviet Union and to its dynamic of developing new nuclear technologies. The potential of the Russian nuclear industry lies in ROSATOM and the Russian defense industry. It is through them that industrial development is capable of drawing on the reindustrialization of post-Soviet Russia.

ROSATOM Atomic Energy State Corporation, the largest company in Russia, has been actively building new power units in the Russian Federation and abroad. Today, the nuclear industry in Russia is a powerful complex of more than 350 enterprises and organizations that employ over 250,000 people. ROSATOM provides 33% of electricity production in the European part of the country. ROSATOM is ranked second in the world in uranium reserves and fourth in terms of uranium creation, and second place in the world in nuclear power generation providing 36% of the world market for uranium enrichment services and 17% of the nuclear fuel market. The volume of uranium production in 2017 amounted to about 3000 tons (in the Russian Federation, not including mining abroad) and electricity generation at the NPPs amounted to 202.868 billion kWh. The share of NPPs in the production of electricity in Russia has risen to 18.9% and in the European part of the country it now exceeds 40%.

The ambitions of ROSATOM are remarkable: the corporation plans to increase its revenues fivefold in the next 16 years—up to $75 billion. The company already controls 25% of the world nuclear market. There are plans for the corporation to enter the Latin American market. ROSATOM is in second place in the world when it comes to controlled uranium reserves having bought a 100% stake in Canada's Uranium One Inc. The company has a globally diversified portfolio of assets in Kazakhstan, the United States, and Australia, and is developing the Mkuju River mine in Tanzania.

Russia's ROSATOM is continuously awarded contracts for the construction of NPPs worldwide. Although China has two state-owned nuclear companies and two local Chinese companies are building 25 reactors, ROSATOM is building 30 reactors there. The aggregate volume of ROSATOM's contracts worldwide is a massive $69.3 billion.

Nuclear Energy in Russia: Recent Developments

In recent years nuclear energy has become one of the most critical sectors of the Russian economy, which is actively developing. The total number of power units under construction in Russia is six. The high quality of the products and services offered is confirmed by its success in acquiring international tracts of land for the development of NPPs outside the country. The portfolio of ROSATOM's foreign orders in 2017 exceeded $130 billion.

Today, Russia is the world leader in the number of power units being built abroad. ROSATOM has signed contracts for the construction of 36 nuclear power units overseas such as the Akkuyu NPP (Turkey), the Belarusian NPP (Belarus), the second stage of the Bushehr NPP (Iran), the Kudankulam NPP (India), the Ruppur NPP (Bangladesh), the second stage of the Tianwan NPP (China), the Hanhikivi-1 NPP (Finland), and the Paks NPP (Hungary).

Dynamic development of the nuclear industry is critical to ensuring the energy independence of Russia and stable growth of the country's economy. ROSATOM's strategy until 2030 suggests that the development of nuclear power will be based on long-term growth and development of new-generation nuclear power technologies including fast-neutron reactors and closed nuclear fuel cycle technology. The export potential of Russian nuclear technologies (particularly, construction of NPPs, uranium enrichment services, and nuclear fuel) is massive.

There are currently (as of 2018) 10 NPPs in Russia operating 35 power units with a total capacity of almost 28 GW. In the overall energy balance of Russia the share of nuclear generation is about 18%. According to the Energy Strategy of Russia for the Period until 2030, electricity production at NPPs should increase between 2.2 and 2.7 times.

In Russian reactors virtually all internals are made of zirconium or rather its alloys E110 and E635. Zirconium has a tiny thermal neutron capture cross-section and a high melting point. Just like lead, zirconium under the influence of radiation inside a reactor has the unpleasant ability to transform from zirconium-92 into radioactive zirconium-93 with a life of 1.53 million years.

If zirconium-90 is used when constructing the fuel assembly, then a very long-lasting material is obtained. Inside the reactor it first transforms into zirconium-91, then to zirconium-92, and only then to the 93rd isotope. The probability a neutron will hit the same atom three times, even with absorption, is minuscule. Therefore, zirconium can be used many times. Thermal neutrons fly through it without hindrance and heat the coolant—not the assembly. The fuel burns more evenly, the fuel assemblies are less deformed, and the same highly active waste is 5–10 times smaller—this is already a considerable saving.

Development of Nuclear Power in Russia

The first circuits of NPP reactors work with water at much lower temperatures than modern coal blocks. Safety considerations dictate that the water temperature in the first course of the NPP is between 250 °C and 350 °C (termed "child"). Water-cooled reactors are now the main types of newly commissioned machines and already form the basis of the modern NPP park.

Before creating materials for the hulls of water-cooled reactors (including testing on "dirty" coal blocks), it was always necessary to choose between radiation and structural safety. Even at temperatures of 300 °C the pressure inside a VVER reactor shell reaches 160 atmospheres.

Unlike water–water reactors, the radioactive vapor from the primary circuit of such reactors goes directly to the turbine. The water pressure in the first circuit of a boiling water reactor is only about 70 atmospheres compared with the 160 atmospheres of a water–water reactor. At this pressure water boils in the core volume at a temperature of 280 °C, which is below the 350 °C of VVER reactors.

Boiling water reactors have several advantages over non-boiling water–water reactors. In boiling water reactors the housing operates at a lower pressure, there is no steam generator in the NPP scheme to transfer heat from the primary circuit to the second circuit in water-cooled reactors, and the daily regulation of such a reactor is much simpler than in the case of a water-powered machine.

However, for stable operation of a boiling water reactor a regime is required in which the mass vapor content in the core does not exceed a specific value. When the collective vapor content is high, operation of the reactor can be unstable. This instability is explained by the fact that steam displaces water from the core and in so doing enlarges the free path of neutrons until the moment of deceleration. If boiling is too rapid, the free path of neutrons enlarges to such a degree that the reactor receives negative reactivity and the power of the reactor begins to fall. Neutrons "fly out" from the core without having any effect on the chain reaction.

In other words, boiling water reactors can be viewed, on the one hand, as rather simple automatic machines whereas, on the other hand, they have much more unstable regimes under which they run the risk of completely "jamming" and bringing about a thermal explosion.

After all is said and done, included in this type of boiling water reactor was the infamous RBMK reactor installed at the Chernobyl NPP. RBMK is the classic boiling water reactor and direct heir of the first Kurchatov reactor at the Obninsk NPP. Lack of understanding of personnel at the station about handling and regulating a boiling water reactor coupled with a mass of abnormal regimes drove the reactor into a state of thermal explosion.

Moreover, because of the low pressures characteristic of boiling water reactors the RBMK at Chernobyl was made without a meaningful body structure; hence, even an explosion of average force could scatter its remains over a vast area. The main problem after the explorion was systematic scraping of graphite from the bitumen roof and cesium from the armor of bulldozers—not reactor shutdown as it cooled down in a week. By the way, the Fukushima NPP was also equipped with boiling water reactors, but they were the American version of the RBMK called the BWR (boiling water reactor). Boiling water reactors and reactors close to the sea under pressure are highly dependent on the coolant and ultimately are unsuccessful from the nuclear physics point of view.

The USSR and then Russia have been investigating sodium reactors for many years starting back in 1959 with the launch in Obninsk of the first fast-neutron reactor, the BR-5, that worked right up to 2002. Its "son," the BOR-60 reactor, is still operational helping to evaluate materials used in the "grandson" of the first BR-5 that is operating now as the BN-600. On the road colloquially called the "Sodium Highway" to sodium fast reactors the USSR and then Russia made errors and incurred losses. Russia lost the BN-350 along this road since it remained in independent Kazakhstan with the break-up of the Soviet Union. Kazakhstan lost its only nuclear reactor as a result of not having personnel skilled enough to maintain such a sophisticated unit in operation.

The BN-800 is being constructed and the BN-1200 is being designed. The Sodium Highway confidently and gradually provided and continues to provide Russia with technologies that are unique to it in the new world where natural uranium-235 costs at least $260 per kilogram and hundreds of thousands of tons of "useless" uranium-238 lies concentrated in Russian enterprises.

Russia plans to build reactors with lead coolant (BREST) and lead–bismuth eutectic (SVBR). In addition to improving the technology of the proven water-cooled WWER reactor the current version the VVER-1000 is planned to be replaced by the VVER-1200 and in the future by the even more powerful and scientifically enhanced VVER-TOI.

The high quality of ROSATOM's products and services offered is confirmed by its success in international competition (especially, tenders for the construction of NPPs outside the country). The portfolio of ROSATOM's foreign orders in 2017 exceeded $130 billion. Today, Russia is the world leader in the number of power units being built abroad because of ROSATOM.

Current State of Nuclear Power in Russia

The nuclear power industry in Russia is a powerful energy complex in which more than 350 enterprises operated in 2018 and at least 250,000 people were employed. It includes four research and production complexes: research institutes scattered throughout the country, engineering enterprises, fuel cycle enterprises, and the defense complex. Development of the industry guarantees the energy independence of the country.

The nuclear power industry continues to develop by actively improving the performance of enterprises. There are currently 10 NPPs in Russia including 35 power units. In 2016 atomic power plants produced 196.37 billion kWh of energy. The share of nuclear energy in the total amount of electricity produced in the country was 18.3%. Russian scientific institutions continue to develop advanced technologies for peaceful use of the atom.

Nuclear energy directly depends on the economic and political world situation. Various events provoke a surge and a fall in interest in the technology of generating electricity from nuclear fuel.

At the same time, new technologies are making the operation of NPPs more secure and preventing emergencies. Russia is not going to curtail its nuclear program since it remains a priority direction for the nation. Russia has vast reserves of energy resources. Russia's powerful fuel and energy complex is the basis of economic development and acts as a robust foundation for domestic and foreign policy. The energy sector ensures the livelihood of people in all areas of the national economy, contributes to consolidation of the constituent entities of the Russian Federation, and ensures the country's economic stability by reliable operation of the energy complex and fuel.

The purpose of the national energy policy is to maximize the efficient use of natural fuel and energy resources, to develop the potential of the energy sector, to ensure economic growth, and to enhance the quality of life of the nation's people. Fuel and energy resources should in the long term allow the structure of power generation to be improved (in particular, increased production at NPPs and fuller use of hydropower). The location of generating facilities determines how electric power industry is developed: either by technical re-equipment of existing thermal power plants, creation of combined cycle power plants, or maximum growth of NPPs. Although solar and other non-traditional sources of electricity can supplement the total energy supply, they cannot substitute for nuclear energy.

Nuclear Weapons of Russia

Russia's Strategic Nuclear Forces (SNF) currently maintain a balance between its combat capabilities and those of US strategic offensive forces. Nuclear weapons, as defined in the updated Military Doctrine of the Russian Federation approved by President Putin on December 25, 2014, "will remain an important factor in preventing nuclear military conflicts and military conflicts with the use of conventional weapons of mass destruction whether large-scale war or regional war" (https://www.offiziere.ch/wp-content/uploads-001/2015/08/Russia-s-2014-Military-Doctrine). This message makes clear the Russian Federation's objective to contain and prevent military conflicts by maintaining nuclear deterrence at a level that is sufficient.

Since the main contribution to Russian nuclear deterrence is made by the SNF (Strategic Nuclear Force) it is interesting to assess their current state and development prospects for the foreseeable future. The SNF of Russia are intended to bring about global nuclear deterrence and include the Strategic Missile Forces, the Strategic Nuclear Forces, the Strategic Nuclear Forces, and the systems providing them. Maintaining the composition, state of combat readiness, mobilization readiness, and preparedness of the SNF of Russia at a level that guarantees damage is unacceptable to any aggressor in any situation is given the highest state priority.

According to the latest exchange of information by parties to the defunct (as of September 1, 2014) START III Treaty, Russia had 911 units of deployed and non-deployed strategic offensive weapons and the United States had 912 units. When it comes to means of delivering these Russia had 528 units with 1643 warheads and the United States had 784 units with 1642 warheads.

Nuclear Legacy of Russia

The presence of nuclear weapons in Russia is both an asset and to a small extent a liability. The chronicle of Russia gaining real and complete control over its nuclear arsenal is replete with good and bad points much like the history of Russia in the early 1990s. The difference is that here the risks went far beyond the borders of the former Soviet Union.

In each of the 15 Soviet republics there was some nuclear activity or nuclear weapons. Moscow was aware of the fact that if atomic weapons remained for an extended period beyond the territory of the Russian Federation, it would be tough to ensure its safety. Tension was fueled by rumors that appeared in the American and Israeli press in early 1992 leaked by "knowledgeable sources in intelligence" that Kazakhstan had allegedly sold Iran one or two nuclear warheads. Although it was clear at the time that these rumors were groundless and had a political connotation, it is unlikely that anyone in Moscow could guarantee such a "fantasy" would not turn into reality.

Russia's nuclear policy called for all the tactical nuclear weapons (TNWs) of the former Soviet Union to be located in Russia, the Ukraine, and Belarus. The Soviet military was able to anticipate the situation and do this before the collapse of the Soviet Union. This significantly reduced the real threat of nuclear proliferation, which would otherwise have become inevitable with the collapse of the Soviet Union.

Withdrawal of TNWs from Azerbaijan was carried out in secrecy from a military airfield. However, a group of civilians from the Popular Front heard about it and blocked the runway in an unsuccessful attempt to stop the planes from taking off. The main problem after the disintegration of the Soviet Union was the deployment of strategic nuclear weapons in the three new states of the Ukraine, Belarus, and Kazakhstan (in addition to Russia). Suffice it to say that the nuclear arsenal at the time of the collapse of the Soviet Union in Kazakhstan exceeded the atomic arsenals of the United Kingdom, France, and China combined.

On December 21, 1991, two weeks after the formation of the Commonwealth of Independent States (CIS) and the day the Commonwealth of Kazakhstan was adopted, four states signed in Alma-Ata (now Almaty) an agreement on joint measures to control nuclear weapons. Then on December 30, 1991 in Minsk the CIS states concluded a deal in which they recognized the need for joint command of strategic forces and the preservation of unified control over nuclear weapons.

On June 6, 1992 nine CIS states (Armenia, Belarus, Kazakhstan, Kyrgyzstan, Moldova, Tajikistan, Turkmenistan, Uzbekistan, and the Ukraine) confirmed that they supported Russia's participation in the Treaty on the Non-Proliferation of Nuclear Weapons (NPT) as a state possessing nuclear weapons and stated that they were ready to join the NPT as non-nuclear states. On this day the question of succession for Russia as a state possessing nuclear weapons was finally resolved.

However, it took another two years for all strategic nuclear weapons to be withdrawn from Belarus, Kazakhstan, and the Ukraine. No problems arose with Belarus and by the end of 1996 there were no nuclear weapons in its territory. Kazakhstan was technologically much the same as the Ukraine and certainly more advanced than Belarus. Despite being capable of independently producing atomic weapons a brief but turbulent discussion arose over whether to become a nuclear state. However, it was severely suppressed by President Nazarbayev who decided to make his country an example of movement toward a nuclear-free world. By the fall of 1996 there were no nuclear weapons in Kazakhstan.

Nuclear Weapons in the World

In January 1994 the Ukraine finally confirmed its non-nuclear status and committed itself to allowing withdrawal of its nuclear weapons (to Russia) in the Trilateral Statement of the Presidents of Russia, the Ukraine, and the United States. At the same time, Kyiv achieved all its goals including Russia's stated obligations to "refrain from economic coercion" and "respect existing borders."

At first, Moscow was convinced that the withdrawal of nuclear weapons from the Ukraine would be resolved in a "brotherly" fashion and without outside interference. As a result, the entire nuclear arsenal of the Soviet Union ended up concentrated in Russia. However, no attempt was made by Yeltsin's Russia to cement cooperation with the CIS states in the field of nuclear energy or nuclear safety as compensation for the withdrawal of nuclear weapons from their territories.

In 1994 Russia ignored a request by Kazakhstan to take several tens of kilograms of enriched uranium. In 1998 Russia refused to accept more than 4 kg of enriched uranium from Georgia. Taking advantage of the hiatus the United States took on the essential role played by Russia in the Soviet Union. The United States began to work systematically with nuclear institutes and specialists from CIS countries. The self-elimination of Russia logically led to its leadership of the CIS being eroded.

The deep economic and social crisis that shook Russian society in the early 1990s almost brought about the collapse of the Russian state, something for which the Russian nuclear industry and the nuclear weapons complex as a whole was completely unprepared.

The nuclear industry could only be saved by export orders. Any romantic ideas about cooperating with a new friend—the United States—quickly struck a number of harsh realities such as the US anti-dumping measures against Russian uranium, sanctions against Glavkosmos for cooperation with India, and the systematic effort of the Clinton administration to undermine Russia. The first Chechen war also caught Russia by surprise. Terrorists looked at nuclear facilities that had no air cover. The situation in the Armed Forces was dire. The Russian atomic shield outwardly somehow held, but knowledgeable people saw that it was broken. The second internal problem was connected with Yeltsin whose decision-making can best be described as a mess. The final result was that Russia became an easy target for the Clinton administration.

Strategic Missile Forces of Russia

Because of its geostrategic position the Soviet Union and then Russia traditionally put the main emphasis on SMF when setting up the SNF.

The leading role played by SMF in the strategic nuclear triad (land-based ICBMs, strategic bombers, and submarine-launched ballistic missiles) is determined not only by their prevalence in terms of the number of carriers (i.e., any device conveying nuclear warheads) deployed (more than 60%) and the number of nuclear warheads (up to 66% of the total amount), but also by their high operational readiness for combat missions (just a few minutes) and the stability of combat control under possible counteraction by an aggressor.

Today the SMF comprise troops in constant combat readiness prepared and willing to carry out combat missions. They represent an independent army in the Armed Forces of the Russian Federation. The Strategic

Missile Force includes three missile armies: the 27th Guards Red Army (headquarters in Vladimir), the 31st Guards Red Army (headquarters in Orenburg), and the 33rd Guards Red Army (headquarters in Omsk). It is estimated that the number of ICBMs deployed is about 350 conveying approximately 1200 nuclear warheads. Up to 96% of this strike group is contained in readiness for immediate application.

Prospects for the Development of Strategic Nuclear Forces

Three main directions have been earmarked for future (until 2020) development of SNF (Strategic Nuclear Forces):

- creation and deployment of new land- and sea-based weapons and delivery systems;
- development and production of diplomatic combat vehicles for advanced combat equipment with specialized combat units and effective means of overcoming missile defenses;
- improvement of the system of centralized combat control of SNF that ensures the delivery of orders to strategic carriers in any situation.

The National Center for Defense Management (NCMC) of the Russian Federation was created in 2014 to evaluate the use of nuclear weapons and to develop an early-warning system and combat management system for the Unified Space System (CEN). It is planned to have an orbital grouping of 10 specialized spacecraft. Modernized command posts will provide control of the orbital cluster, reception, processing, and transmission of private information in automatic mode making it possible to improve the timeliness of detection of a missile attack on Russia, the reliability and efficiency of bringing orders from centralized combat control to the SNF in response to a nuclear attack on Russia.

The fact that Russia will not be able to achieve parity with the United States in the number of strategic offensive weapons by 2019 will no longer be critical to national security. Based on the assessments made by the Russian expert community, it seems acceptable to have 530–570 carriers deployed conveying about 1550 warheads in the SNF of Russia by this date.

There is no doubt that by 2018 the US SNF had 700 carriers deployed conveying 1550 warheads. What is more, the Americans will have a so-called return potential when it comes to strategic offensive weapons. However, this superiority the Americans have in terms of their return potential will not have a significant impact on Russia's nuclear deterrence potential. By 2021 Russia is expected to have 700 carriers deployed in any case. That will ensure maintaining Russia's SNF at a level that not only solves the problem of inflicting damage that is unacceptable to any aggressor, but also strategically deterring any nation from unleashing a war against Russia and its allies in any military–strategic situation.

The American Pressure on Russia Regarding Iran and India

By signing an agreement for the construction of the Bushehr NPP in Iran in 1992 Russia felt that it was gradually getting entangled. On the one hand, desperately needed revenue would come from the first Russian nuclear contract in the Middle East. This would lead to new contracts not only in the atomic sphere, but also in hydrocarbon power engineering and military–technical cooperation. On the other hand, Russia's Foreign Intelligence Service said in 1993 that Iran was implementing a "program of military applied nuclear research."

Back in 1995 Moscow proposed considering Russian–Iranian nuclear cooperation as "a kind of training ground on which the possibility and necessity of the fulfillment by the member state of the nuclear club of its obligations under Article IV of the NPT, in which the parties to the Treaty should promote equitable, non-discriminatory cooperation in the field of peaceful nuclear energy, but at the same time not to allow conditions for the proliferation of atomic weapons."

US pressure on Russia in connection with its cooperation with India was another painful lesson. In 1992 Senator Al Gore during the election campaign initiated a ban on Russia supplying cryogenic rocket engines to

India. Moreover, although Russia was not at the time a member of the Missile Technology Control Regime (MTCR) and had no such obligations, it was forced to comply with US requirements.

In 2000 Russia decided to supply 58 tons of uranium dioxide to India at an NPP in Tarapur to ensure safe operation of the plant. Washington regarded this as a "serious threat" to the non-proliferation regime. Although Russia did not give in, the irony of the situation—and the lesson—is that in a few years it will be the United States that initiates the lifting of restrictions on nuclear trade with India imposed by the Nuclear Suppliers Group.

Implications for US Nuclear Deterrence and Missile Defense

Russia has modernized almost 60% of its strategic forces. It has declared that it will "continue a massive program of nuclear rearmament, deploying modern ICBMs on land and sea, and modernizing the strategic bomber force. This new strategy allows for the first use of nuclear weapons in limited conventional warfare" (https://www.swp-berlin.org/fileadmin/contents/products/research_papers/2009_RP12_kle_ks; *The Priority Tasks of the Development of the Armed Forces of the Russian Federation*, The Defense Ministry of the Russian Federation, 2003, p. 70).

Vice-Admiral (Retd.) Robert Monroe, former Director of the Defense Nuclear Agency in the United States, has recently warned that Russia is now 20 years ahead of the U.S. in these weapons. However, this may simply be a ploy of the Pentagon to get more funds.

Currently Russia is developing two new systems: the improved Moscow Anti-Ballistic Missile system and the S-500, both of which are effective against any ICBMs (Interfax, December 19, 2012). Deputy Defense Minister Yuri Borisov said the S-500 "could destroy aerodynamic and ballistic targets of all types…" (TASS, March 26, 2015). He said that the S-500 could intercept "low-orbital satellites and space weapons" and "intercontinental ballistic missiles in the terminal phase of the trajectory and, within definite limits, in the midcourse sector" (Interfax-AVN, Moscow, April

24, 2013). The S-500 is more capable than the new Russian S-400 against stealth aircraft.

In 2018 Russia announced that it was reviving the military program set up during the Cold War to provide the general population with concrete shelters against nuclear attack. That was in response to US plans that included withdrawing from the agreement on intermediate-range and short-range missiles (SMRBMs) and thinking of launching a preventive (first and unannounced) disarming massive nuclear missile strike against Russia's SNF with intermediate-range nuclear weapons (NWM).

We live in a world based on nuclear capabilities. Russia is indeed one of the most significant nuclear powers and is improving its strike systems. Some are already in service and others will be delivered by 2020. In short, the Russian Federation is actively developing systems similar to its Iskander and Caliber missile systems. Although both have conventional warheads, they can also carry nuclear weapons. In addition, new atomic military systems have been set up in Russia such as the Avangard, the Sarmat, the Poseidon, the Kinzhal, the Bulava (on nuclear submarines), the Caliber, the Rubezh, and the Yars, all of which automatically operate the Perimeter Dead Hand system. Thus, the strategy of pre-emptive strategic deterrence allows Russia not to increase defense expenditures by developing nuclear weapons and not to participate in the arms race in response. According to Dave Majumdar, an expert from the US Institute of Defense and Strategic Studies, "Russia's SNF rearmament programs are at an advanced stage. Currently, Russia already has a solid advantage over the U.S. in terms of quality and diversity of delivery systems, and can reliably ensure the strategic effectiveness of its nuclear forces shortly" (https://inosmi.ru/politic/20180622/242556917.html).

However, although Russia's military doctrine in the use of nuclear weapons does not include a preventive strike, it does include a reciprocal response to attack. This means that Russia is ready to use nuclear weapons only when a potential aggressor strikes Russian territory. The entire history of Russia shows that it has never been the initiator of such catastrophes. However, as the President of Russia said in 2018 at a meeting of the Valdai International Club, "The aggressor should know that we, as martyrs, will go to heaven, and the aggressor will simply die before he has time to repent" (https://nsarchive2.gwu.edu/). Thus, the President of Russia declared that the nation's readiness to deploy new super-technological strategic systems was only a "pre-emptive deterrent" to the arms race and nuclear war. That once again reminds the world of the irreversibility of the end of the 500-year domination of the West worldwide. To reverse the course of history, the West is essentially starting a new Cold War to regain positions lost over the past decades. However, the world is different now and the result will be different.

12

Japan as Nuclear Power

Japan is the only victim of nuclear destruction. Destruction of Hiroshima and Nagasaki by the US nuclear bomb attack in 1945 forced Japan to abandon nuclear weapons forever, but the recent rise of China is forcing Japan to change its passivity. However, since Japan currently is under the US nuclear umbrella, there will be no immediate change in its nuclear weapons policy.

The Peaceful Atom of Japan

It all began in 1954. Approximately in this period, many countries of the world were engaged in developments in the field of nuclear energy. Japan decided not to stand aside, allocated millions of Yen from the budget for the development of the industry, regulated the use of nuclear technology for peaceful purposes, and two years later, the Atomic Energy Commission was established. At the same time, several other scientific organizations were created to study nuclear energy.

D. Basu and V. W. Miroshnik, *The Political Economy of Nuclear Energy*,
https://doi.org/10.1007/978-3-030-27029-2_12

The first Japanese nuclear power plant is Tokai-1 Nuclear Power Plant, located on the island of Honshu, to Tokyo from this place is only 130 kilometers. The nuclear power plants began to be built already in 1961, the Magnox 160 MW gas-cooled reactor was imported from the UK. The British company The General Electric Company (GEC) completed all construction work by 1966.

Tokay-1 NPP is known not only for the fact that it was the first in Japan and the opening closed but also because, on September 30, 1999, there was one of the most severe accidents in the atomic history of the country. It happened at a small radiochemical plant of the company JCO. It was involved in the processing of isotope enriched uranium hexafluoride in uranium dioxide, from which fuel was subsequently produced for some commercial nuclear power plants in Japan.

This plant began operating in 1988; annually processing up to 3 tons of uranium. The incident occurred during the uranium purification procedure. It would seem that the work of the plant had to be controlled, and the process itself had to be strictly regulated. Tokay Nuclear Power Plant worked until March 31, 1998. The reactor was finally dismantled in 2011.

The Tokai-2 nuclear power plant with the BWR-5 type reactor, 1100 MW, was built in 1971. General Electric also developed this power unit with a boiling water reactor. The power unit was the first in Japan, whose capacity exceeded 1000 MW. The station became active in 1978.

The nuclear power plant was automatically shut down after the accident at the Fukushima-1 nuclear power plant in March 2011, but the radioactive water treatment plants have partly continued their work. Three years later, it was decided that the second unit of the Tokai nuclear power plant was still needed, and the station's management applied for a restart. In 2014, in total, 18 applications were submitted for the restart of reactors, and Tokai-2 became the oldest power unit that participated in this.

Meanwhile, before the nuclear accident, which shook the whole world, the world's largest nuclear power plant was located in Japan in the city of Kashiwazaki—Kashiwazaki-Kariva nuclear power plant. There are five BWR reactors and two ABWR reactors. The total capacity of the station is 8212 MW. For comparison, only one of these nuclear power plants is twice as powerful as all those operating in the Czech Republic, and four times that of Hungarian nuclear power plants. In addition to the enormous capacity, the station is also remarkable for the fact that it was here that the first reactors of the type ABWR (boiling nuclear reactors of the third generation) were built. Because of frequent earthquakes, nuclear power plants were often turned off for inspections and restoration works. Now this station, like many others, is turned off.

Accident at the Fukushima-1 Nuclear Power Plant

In general, Japan is a particular example of a country whose investment natural conditions primarily determine climate in the nuclear industry. It is impossible to ignore other factors that affect the current sectoral investment climate. These include, first, subjectivism and populism in the assessment of nuclear safety issues, which have grown significantly after the accident in Fukushima. Secondly, the traditionally cumbersome bureaucratic procedure for controlling nuclear safety, which reduces the economic efficiency of capacity utilization (it has always been lower in Japan than in most developed countries). Thirdly, the significant role of municipal authorities in approving nuclear projects: On the one hand, they are subject to public opinion, on the other—many municipalities directly decide the fate of operating and under construction NPPs, having great powers of coordination, and in some cases being actual co-owners of stations.

All these factors lead to a noticeable regress in Japan's nuclear power industry after the Fukushima accident: Most of the blocks were idle for several years, and several new capacity projects were frozen. And this despite the fact that nuclear generation in the foreseeable future is the best way to increase the energy independence of Japan, a country that is highly dependent on foreign energy supplies (distributed Renewable Energy Sources [RES] has until recently played no significant role in Japan, and the planned buildup of this sector, taking into account the need increased reserve capacity, stretched for decades).

In this regard, it is appropriate to compare Japan with South Korea, which, having a comparable dependence on fuel imports made an unambiguous bet on expanding nuclear power with the most vigorous patronage of the state. However, the main difference of the country of morning freshness from the Land of the Rising Sun is the significantly less seismic activity of the areas where Korean nuclear power plants are located and the probability of extreme tsunamis is much less likely (opposing points of view were expressed but not supported in the expert community). Accordingly, there is no similarity to the negative Japanese experience of emergencies at nuclear power plants caused by natural disasters.

The Atomic Policy of Japan After Fukushima

Nuclear energy is a whole set of factors related to economics, politics, energy itself, ecology, psychology, social sphere, and much more. No matter how high were the hopes of the "green" on RES, and how actively the anti-nuclear lobby did not push its position, nuclear power proves a number of its advantages over other energy sources.

For many countries, this is simply a vital source of energy, as well as political and energy independence. By the way, according to many experts, including from ROSATOM, the Russian nuclear energy company, the future of world energy is not a choice between alternative energy and nuclear and traditional sources, but a symbiosis of various types of energy carriers. Also, with "green" energy it is not so simple: Despite the reduction in its cost, it continues to be expensive and poorly developed in some countries only because of economic and climatic conditions. At the same time, without nuclear power, which frightened humanity with several severe catastrophes, it would be impossible to imagine the future of at least fifty years.

Even though this country became the only one severely affected by nuclear weapons, in the 70s of the last century, it was nuclear energy that became the national strategic priority. Before the accident at the Fukushima-1 nuclear power plant, a total of 54 reactors operated in the country. This means that Japan was the third largest in the world in terms of the number of reactors, and 30% of the total energy balance was accounted for by nuclear power plants. The catastrophe made not only Japan but also several other countries, especially Germany, to forcefully abandon nuclear energy. True, it did not last for several years, as the new government of Shinzo Abe decided that without nuclear generation, the economy can go nowhere, and took up the revival of nuclear power. After all, it is in severe energy and economic situation due to the lack of the necessary volumes of its energy resources. It has to depend on imports of oil, coal, and gas.

After the catastrophe at the Fukushima-1 nuclear power plant, Japanese society began to react negatively to nuclear power. The reason for this is apparent. However, it was this severe accident that contributed to the fact that the stations around the world were adopting more stringent standards of work and safety, and also tightened the requirements for the new stations being created. In this sense, the Tokay-1 NPP after Fukushima has become much safer. The Japanese, considering the terrible experience, erected around the station a wall 18 meters high to protect the nuclear power plant in the event of a tsunami. Also, the ventilation system was updated, and 18.5 thousand meters of cables were covered with fireproof material.

The Japanese, according to the expert, were once subjected to a nuclear attack, are extremely afraid, perhaps more than the population in other countries. Therefore, it is possible that this could be one of the reasons that after the accident at the station, the necessary operational measures were not taken to normalize the situation for a long time. In particular, after the crash, almost all the station's employees were evacuated, and there were no people who could connect the backup generators. As already mentioned, there are 54 nuclear reactors in Japan, after Fukushima, all reactors were suspended for a year for stress tests and the construction of additional protective equipment. In 2012, Yoshihiko Noda, the then Prime Minister of the country, announced that the government had headed for complete abandonment of nuclear energy by 2040 after the existing reactors had reached the age of 40 years. This age of reactors is the last license period for their operation. In 2015, it was decided to dismantle five power units for four Japanese nuclear power plants: Tsuruga, Mihama, Gankai, and Shimane, since their age reached 40 years.

However, in 2012, Shinzo Abe, who came to power, declared that Japan could not do without nuclear power. In the same year, new Nuclear Regulations Authority standards were established to regulate nuclear safety. It was decided that so far, at least three reactors will start working again on these standards.

Two years later, the government issued its first strategic energy plan after the Fukushima accident. The Ministry of Economy, Trade, and Industry in 2015 published a long-term project, which by 2030 envisages increasing the share of nuclear generation in the country's total energy balance to 20–22% of electricity.

If before the "Fukushima" anti-nuclear movements existed in the country, but the Japanese still generally supported pragmatic development from the practical reasons, then, according to the results of the polls in 2015, 47.9% of the respondents said that the nuclear power industry should be gradually canceled, 14.8% favored a complete and immediate abandonment of nuclear power plants. Only 10.1% of respondents said that the use of nuclear energy should be preserved, and only 1.7% said that its volumes should be increased. For example, on one of the anniversaries of the accident at the Fukushima, more than 9000 people went to a rally with posters "Stop the work of the nuclear power plant," "Goodbye, the atom." Also, some political forces of the country support this position of citizens.

According to IAEA data from 2016, the share of nuclear energy in the country's total energy balance was only 2.15%. Over the past year, a total of 810,260.00 GWh was produced, of which 17,453.00 GWh was spent on nuclear power plants. For comparison: In 1997, the share of nuclear power plants reached 35%—this was the peak of the atomic energy in the country. Until 2011, this indicator fluctuated slightly but still did not fall below 25%. What is the nuclear path of this Asian country? What does nuclear energy mean to it? What is the specificity of this energy sector in the Land of the Rising Sun?

By the degree of development of competition in the industry, it is at the level of countries with the most liberal environment in nuclear energy. Today, Japan no longer plans to abandon nuclear generation (as the previous government assumed) and does not impose draconian fiscal measures. However, extremely high safety requirements, caused by the bitter experience of natural disasters and the objectively complex conditions for the construction of nuclear power plants, lead to increased costs for investors, which sometimes reach a prohibitive level for investment.

The objective reasons for "reinsurance" include, first of all, the extremely complex seismic conditions and the danger of extreme tsunamis: All the sites of operating and under construction Japanese nuclear power plants are in zones of seismic activity, similar to which most countries of the world are considered unacceptable for the placement of nuclear power facilities. After a series of natural cataclysms—some of them caused significant damage (Kashiwazaki-Kariva nuclear power plant) or severe accidents (Fukushima-1 nuclear power plant)—Japanese supervisors tightened the already harsh security requirements.

For example, the maximum horizontal acceleration of a safe stop (one of the main parameters of seismic stability) for most nuclear power units in the country is set at 0.45–0.80 g, and for some exceeds 1–2 g. Seismic stability is not a feature in nuclear power plants built in other countries of the world possess seismic stability, which fits into the first range; the second level is unprecedented.

Measures to strengthen the resistance to earthquakes and tsunamis, which originate from the "post-focus" installations of the supervisory authority of Japan, require enormous investments—in addition to those that have already been implemented beyond the original design. Under these conditions, a number of nuclear power plant owners prefer to close nuclear power blocks, since modernization, according to their calculations, will not pay off; the necessary amounts are hundreds of millions and billions of dollars, which is economically impractical for blocks of medium capacity of old structures (less than 500–700 MW). So, in 2015, the owners closed one power unit at Shimane, Genkai, Tsuruga, and two groups at the Mihama NPP.

From Words to Actions

In 2015, after two years of downtime, the first power unit of the Sendai nuclear power plant with a capacity of 846 MW was launched in the

Japanese city of Satsuma Sendai on the island of Kyushu. Later the company-operator Kyushu Electric Power Co. (KEPCO) reported that the power unit is connected to the leading network and generates electricity. Then the second reactor of the station began work. Initially, the two reactors of the Sendai nuclear power plant were to be restarted in early 2015, but in the design documents that were needed to restart the plant, the drawbacks in the drawings and seismic data were identified. Still, the reactors were recognized as meeting the new safety criteria for nuclear power plants in 2014. The residents were against restarting the station, they demanded to abandon the nuclear power plant, but the court rejected their lawsuit, finding the fears of citizens in vain.

Until the moment of stopping the Sendai NPP in connection with the accident at the Fukushima-1 NPP, the total capacity of the station was 1780 MW. Both installed reactors belonged to the PWR type produced by Mitsubishi Heavy Industries. In general, this nuclear power plant was the first to be launched after the disaster. The government made this decision of Kagoshima Prefecture. Such a measure was due to the difficult economic situation and the high cost of traditional energy sources, in particular, liquefied gas.

When Abe demanded to restart another 21 reactors began to be considered. Despite this, in many Japanese cities, the intention to restart the nuclear power plant provokes discontent among the local population. For example, the Fukui Prefectural Court prohibited the restart of the Takahama nuclear power plant.

Ikata NPP is the only nuclear power plant on the island of Shikoku, the fourth by area and population in Japan. The station was launched in 1977. The nuclear power plant consists of three power units with reactors of PWR type manufactured by Mitsubishi Heavy Industries. This station, as a result of inspections and technical improvements, proved its safety; it is mainly protected from earthquakes and tsunamis, which the population is so afraid of after the disaster. On May 21, 2015, the Japan Atomic Energy Control Committee gave the go-ahead for the restart of the third 890 MW power unit built in 1994. That is the only power unit in the country that uses MOX fuel, which is based on a mixture of uranium and plutonium.

In general, after the "Fukushima" most Japanese nuclear power plants—in a state of cold freezing. Two reactors were under construction, but after the accident their construction was suspended. Currently, of the 42 operating reactors in Japan, three have been restarted: on the blocks 1 and 2 of the Sendai Nuclear Power Plant and at the No. 3 of the Ikata Nuclear Power Plant by Shikoku Electric Power.

A few more nuclear power plants may soon restart. The first—NPP “OI,” located in the prefecture of Fukui. According to the estimates of the Japanese nuclear regulator, the station corresponds to post-fusil security standards. That is primarily about restarting the third and fourth power units of the station. In April 2017, discussions will be held with the public about this, with successful completion of the reactors can again earn this fall. The second station, which could soon be restarted, is Genkai NPP in Saga Prefecture. Therefore, in February, most members of the municipal council of the town of Genkai voted for a restart, in March the mayor of the city took the same decision. The region views nuclear power as the primary energy source, so restarting these four power units is of great importance for Japan.

What Does a Nuclear Power Plant (NPP) Mean for Japan?

In the twentieth century, a confidently taken course toward the development of nuclear power in Japan gave fruit. For example, by the end of the 1970s, the industry in the country had built up a significant amount of its nuclear capacity. Approximately at the same time, Japan, like Germany, France, Sweden, and Switzerland, began to knock on the conditional club of the first suppliers of nuclear technologies (the USSR, the USA, and Canada). The leading Japanese suppliers—Hitachi, Toshiba, Mitsubishi Group, who for the first time focused exclusively on the domestic market, exported only specific components and equipment. Later in Japan, they created their versions of the BWR reactor; the same happened in Germany. Japanese and German companies have gradually squeezed out General Electric both from their domestic markets and from the nuclear markets of other countries.

Then, since 1978, General Electric, in cooperation with Toshiba, Hitachi, ASEA Atom and Italian Ansaldo, has begun developing a third-generation ABWR reactor with a capacity of over 1300 MW. The mid–80s completed the project by General Electric, Toshiba, and Hitachi with the financial support of six Japanese energy companies led by TEPCO. Japanese companies did not immediately seek to win foreign markets alone; they acted in alliances or acted as intermediaries. In commercial nuclear power engineering in the country, the company specialized in boiling reactor plants for about 50 years, again not in itself, but cooperation with General Electric.

Also, in Japan, assemblies of fast neutron stations were developed. The Atomic Energy Agency of Japan (JAEA), which was formed from the merger of several government agencies, created the first critical assembly on fast neutrons FCA. It was launched in 1967. Ten years later, the experimental fast reactor "Gioio" was constructed, and in 1996—an experimental fast-neutron reactor "Monju." In general, Japanese nuclear engineers had many successful projects, the more difficult it is to imagine that Japan could once and for all forget about its achievements, shut down all its reactors by 2030, as it was supposed after Fukushima, and put an end to the plans to introduce their technologies in other countries.

Also, even the accident at Fukushima-1 could not wholly shake Japan's export plans. In December 2011, the House of Representatives of the Parliament of Japan ratified the agreement with Russia, Jordan, South Korea, and Vietnam on the peaceful use of atomic energy. Despite the concern of the Japanese population, Yoshihiko Noda, who at that time held the post of the prime minister of the country, said in his speech to the lower house of parliament:

After the disaster, it's been six years already, and Japan has at least slightly moved away from this horrific event, but, of course, with many reservations. In July 2016, representatives of the Japanese financial-industrial group Hitachi Ltd. announced the cooperation of their subsidiary British company Horizon Nuclear Power with the Japanese nuclear power company Japan Atomic Power Company (JAPC) for the construction of nuclear power plants in the UK. According to Hitachi's plans, by 2020 the companies will build in the United Kingdom 4–6 advanced boiling water-cooled reactors to replace the non-operating British nuclear power plants "Wolfe" and "Oldbury." Japan will supply 40% of all necessary details for the construction of stations.

It was noted that Hitachi Ltd's plan to cooperate with JAPC in the construction and operation of British nuclear power plants is a significant moment for Tokyo because the Japanese have long wanted to start exporting their atomic technologies. As REGNUM reported, in December 2016, Japan and Great Britain signed a memorandum of understanding on joint work in the nuclear field. Moreover, the Japan Bank for International Cooperation and the Development Bank of Japan can provide financial assistance up to $8 billion 500 million for the construction of nuclear power plants in the UK.

Also, Japan had signed an agreement to develop nuclear energy with India. This document allows Tokyo to supply its nuclear technology to this country within the framework of civilian energy, which is a kind of a new precedent for Japan. India did not sign the Treaty on the Non-Proliferation of Weapons, and the Japanese side is an active propagandist of the NPT and with such a country for the first time entered into such an agreement. Japan wants to export its reactors abroad, and India has plans to increase its nuclear power capacity by about ten times and attracts many world players for this by 2032.

In March 2017, the governments of Japan and France agreed to develop cooperation in the field of nuclear energy. In France, the share of nuclear generation is 72.28% in the total energy balance. Japanese Mitsubishi Heavy Industries and Japan Nuclear Fuel acquire a 5% stake in the nuclear fuel processing plant, which will soon be built by the French Areva.

In general, the attitude of the Japanese authorities to nuclear energy as an energy source within the country, as an export commodity, is purely pragmatic. As long as without a "peaceful atom," it is impossible to provide energy for the stable development of the country's economy and maintain a high level of household comfort of citizens.

Japan is making a severe bet on alternative energy. By March 2017 the country plans to add 300 MW of wind power. As noted in the study of the Japan Wind Energy Association, this is enough to supply more than 100,000 average houses. The current level is 3.4 GW (1.5% of the total installed capacity of the country's generators), and by 2030 authorities want to increase these figures by a factor of ten. So far, 90% of the available "green" capacity falls on solar energy, but it is also planned to develop wind power also. Today, electricity from the latter is costly, so from October 1, 2017, the Ministry of Economy, Trade, and Industry of Tokyo (METI) will reduce preferential tariffs for large wind energy projects from 22 yen to 21 yen per 1 kWh. This trend is understandable, but it does not run counter to the government's plans for the development of nuclear energy.

For the Land of the Rising Sun, the decisive factor for the use of nuclear energy is the almost complete absence of its energy resources. Japan is entirely dependent on energy imports. It is the world's largest importer of liquefied natural gas, second in terms of coal purchases and the third largest in terms of oil imports. In connection with the abandonment of nuclear power, there was a decline in industrial production. For comparison: If in 2010, Japan bought 94.5 billion cubic meters of gas, then in 2013—119 billion cubic meters. The leading LNG suppliers to this country are Australia (about 19%), Qatar (18%), Malaysia (17%), Russia (10%), and Brunei (7%).

Japan spent $270 billion for the purchase of coal, oil, and LNG three years after the disaster at Fukushima, which is 58% higher than the costs that could have been generated by the operation of the nuclear power plant. As a result, the country faced a trade balance deficit for the first time in 30 years. The figures speak for themselves: in 2010, the budget surplus is $65 billion, in 2013, there is already a deficit of $112 billion. At the same time, domestic energy prices have raised very much; the population has started paying 20% more. From its environmental obligations to reduce greenhouse gas emissions, Japan was forced to abandon temporarily, as the nuclear power plants were closed, and oil and coal production, on the contrary, increased. The country was forced to plan the construction of new coal-fired power plants. At the same time, the amount of greenhouse gas emissions in 2013 reached a record.

2011 was a critical year for nuclear energy. After the earthquake in Japan and the accident at the Fukushima nuclear power plant, during the year, all 54 atomic blocks of Japan were stopped. Raw materials for manufacturing fuel for Japanese atomic power plants were supplied from the United States and Europe. The international Franco-Japanese alliance AREVA—Mitsubishi Heavy Industries supplied fuel to half the reactors, to the second US-Japanese transnational corporation Global Nuclear Fuel (GNF)

The Japanese disaster, although it did not directly affect the interests of Westinghouse, was the turning point when the company moved to its logical end. The struggle for the East European market for Westinghouse is not only the promotion of American interests in this territory and the elimination of Russia's influence, but it is also a solution to the problem of survival.

The fact is that the NPP market is not flexible enough, but it has its tendencies. Today Westinghouse supplies its fuel for seven blocks in Belgium, one block in the Netherlands, two blocks in Finland, two blocks in South Africa (all in half with AREVA), 14 blocks in the UK, one in Slovenia, 48 in the United States, two in Taiwan, two in Switzerland, and seven blocks in Sweden.

At the same time, the Belgian nuclear power plants will be closed by 2025, and the process of stopping the blocks will begin very shortly. All the old British reactors are likely to be shut down in the 2020s, and instead, Europeans, led by the French, will build new ones. Moreover, then Westinghouse has no chances to get contracts.

Given all these circumstances, it is evident that nuclear power is necessary for Japan because of energy, economic, and political problems. Also, the disaster at Fukushima-1, although it undermined the Japanese's confidence in nuclear power plants, still contributed to the development of new, more sophisticated, and harsh rules of construction and operation. Ultimately, nuclear power is almost the only instrument of Japan for energy independence and the development of its economy. Nevertheless, according to experts, as a result, the Land of the Rising Sun can still abandon the nuclear power plant. Therefore, in what direction can Japan eventually bend, it is difficult to say, but for now, it chooses the "peaceful atom" as the basis.

Appendix

US Design on India's Nuclear Programme

Russia after India's nuclear tests in 1974 was the only country for India as a source of materials for the nuclear energy including weapons and nuclear power plants. The only exceptions was the USA, which has continued to supply fuels for the Tarapur nuclear plant for some years and China, who has supplied fuels for the Tarapur plant after the refusal of the US to do so. Russia so far has supplied India heavy water plants, reprocessing plants, Fast Breeder Reactors (FBR) and two fresh nuclear power plants with the excuse that the contracts for these were signed between India and the Soviet Union, which Russia has to oblige.

Recently USA had suggested that if India would purchase nuclear power plants not from Russia but from the US, these restrictions can be relaxed in case India can separate out the civilians power plants from those which can be utilized by the Indian defense establishments. Both of these steps require massive financial investments. The purchase of nuclear power plants from USA would mean paying possibly ten times more than the cost of corresponding Russian nuclear power plants. There would not be any technology transfer either.

The nuclear offer of the US was a process of surrender for India regarding its nuclear energy and weapons programme. If India buys nuclear power plants from the USA, these would be under full-scale inspection of the IAEA, thus India cannot divert anything from these plants for the defense services. India will not get the reprocessing plant, without which it cannot manufacture nuclear weapons. Thus, gradually India nuclear weapons programme will disappear. India's efforts to develop missiles and rockets would be affected adversely too.

Thus, India would be much less powerful than Pakistan in every aspect, as Pakistan has no such constraints. It would continue to receive both advanced nuclear weapons and missiles from China in future as it had received in the past with the full knowledge of the USA since the days of President Reagan.

India's Position on Nuclear Plants

The realistic option for India is the Kalpakkam FBR, built by Russia to provide India enriched uranium for the nuclear weapons. However, that route also can be closed by the USA who is increasing pressure on Russia through the Nuclear Suppliers Group (NSG) of 44 nations. Unless India will abandon its nuclear weapons, it cannot be a member of the NSG either. To override the objections of the NSG, Russia has offered India a floating nuclear reactor, which can be placed near India's shore. However, India is so far reluctant to accept it, as it would certainly annoy the USA.

The Reality Regarding Nuclear Drama

The Vienna Treaty with NSG has clearly pointed out the following that would cripple India's nuclear weapons programme. The Treaty says:

1. "Participating Governments may transfer trigger list items and/or related technology to India for peaceful purposes and for use in IAEA (International Atomic Energy Agency) safeguarded civil nuclear facilities."
2. "Participating Governments may transfer nuclear-related dual-use equipment, materials, software, and related technology to India for peaceful purposes and for use in civil nuclear facilities under IAEA safeguards."
3. "At each Plenary, Participating Governments shall notify each other of approved transfers to India. Participating Governments are also invited to exchange information, including about their own bilateral agreements with India."

Thus, only for the facilities under the control of IAEA India can import fuel, materials and technology. It cannot import these even from Russia, as Russia is a member of NSG and has to consult NSG to export any nuclear materials or technology to India particularly anything related to India's Nuclear Weapons Programme. This was made clear in the Senate bill presented when the discussion on the 123 treaty with USA is in progress.

That would practically mean FBR and Reprocessing plants, being built by Russia, would be under the IAEA controls, otherwise Russia cannot supply any materials or technology for these from now on. Thus, India's nuclear weapons programme or the reactors in the military sector would be lame ducks for the foreseeable future unless and until Russia would come out of NSG.

123 Deal puts further restrictions. The Bush administration's January 2008 letter to the House Foreign Affairs Committee, made public recently, brings out the following.

> USA has given no legally binding fuel-supply assurance of any kind and there is no US consent to stockpiling of fuel reserves. India will not be allowed to build such stocks to avoid if US re-impose sanctions. US civil nuclear cooperation is explicitly prohibiting further nuclear tests by India even if warranted by Indian national security concerns. All cooperation will cease immediately if India conducts a test.

The US has retained the right to suspend or terminate supplies at its own discretion.

The 123 Agreement has granted India no right to take corrective measures in case of any fuel-supply disruption. Rather, India's obligations are legally irrevocable and perpetual.

The Bush administration's letter states that the 123 Agreement fully conforms to the Hyde Act provisions. "US government will not assist India in the design, construction or operation of sensitive nuclear technologies." Under the 123 Agreement, India has agreed to forego reprocessing until it has, in future, won a separate agreement.

123 Agreement Article 5(2) that, "Sensitive nuclear technology, heavy water production technology, sensitive nuclear facilities, heavy water production facilities and major critical components of such facilities may be transferred under this Agreement pursuant to an amendment to this Agreement." The Bush administration's letter to Congress states that the US government had no plan to seek to amend the deal to allow any sensitive transfers.

Already the US senate has imposed a new clause in the Hyde Act that in future national security organizations of the USA, which means

CIA and FBI, would now collaborate with India regarding nuclear non-proliferation. This in effect would imply that US organizations would make sure India will not be able to gain any advantage to use its nuclear facilities to create nuclear weapons.

Section 104(d) (2) of the Hyde Act is related to the supply of nuclear fuel to the plants in India, which would be used to produce nuclear weapons, by using end-use monitoring of spent fuel by the IAEA and the US organizations. There are provisions in the legislation, which would put a cap on fissile material production. These would effectively end India's nuclear weapons programme.

Alternative Was Available

India was not a "Pariah" in the world of nuclear energy since 1974, but India has become nearly self sufficient due to the help from the USSR and Russia. Russia, after India's nuclear tests in 1974, was practically the only country for India as a source of materials for the nuclear energy industry including both nuclear weapons and nuclear power plants. Russia along with the former Soviet Union so far has supplied India heavy water plants, reprocessing plants, FBR and two fresh nuclear power plants with the excuse that the contracts for these were signed between India and the Soviet Union in 1985, which Russia needs to oblige.

USSR has started for India the construction of the first FBTR (Fast Breeder Test Reactor) of capacity of 40 MWt (million watts thermal) in Kalpakkam in 1985. In 2008 Russia has started the construction of a Prototype FBR of capacity 500 MWE (million watts electrical) in Kalpakkam. Associated Reprocessing plants in Kalpakkam were first built by USSR in 1985 and then in 1998 by Russia.

Without the nuclear deal with NSG and USA India could be able to maintain its nuclear plants by using reprocessed plutonium as a fuel in the FBRs and using its own uranium in the conventional plants. India could continue to get both onshore and offshore nuclear plants from Russia, as it would honour the Indo-Soviet Treaty of 1985. This is exactly what former President Putin and former Prime Minister Fradkov have suggested during their visits to India, but India was not interested. In that case India would be at liberty to test and develop nuclear weapons any time it likes without any restrictions using plutonium from its FBRs and enriched Uranium from other nuclear plants. Indo-US Nuclear Deal and the Vienna Treaty with NSG on the other hand have increased both real and perceived restrictions on India that would in reality destroy any credible nuclear deterrent for India against possible attacks from either China or Pakistan.

Section 103 of the Hyde Act suggests that the US would oppose development of a capability to produce nuclear weapons by any non-nuclear weapon state within or outside the Nuclear Non-Proliferation Treaty regime. The section requires the US to work with the 45-nation NSG to further restrict transfers of equipment and technologies related to uranium enrichment, reprocessing of spent nuclear fuel and production of heavy water to all countries, including India. The legislation also requires the US government to seek to prevent transfer of these equipment and technologies from other members of NSG or from any other source. Section 104(d)(2) stipulates that transfers to India cannot begin without the NSG guidelines. Also, there are provisions in the legislation, which would put a cap on fissile material production. The US Senate Bill can apply Hyde Act worldwide against India even when India would buy anything nuclear from any other countries within NSG.

13

India as a Nuclear Power

History of Nuclear Development in India

The nuclear energy sector is the fifth-largest source of electricity in India after coal, gas, hydroelectricity, and wind power, and its contribution is increasing. As of March 2018, India has already twenty-two nuclear reactors in operation in seven nuclear power plants of various types, with a total installed capacity of 6780 MW. Nuclear energy sector produced a total of 35 TWh of electricity in 2017. Six more reactors are under construction with a combined generation capacity of 4300 MW.

Population growth and industrialization continue to be significant sources of growth for the Indian economy. India is among the top ten countries in the world by the number of nuclear reactors. According to the IAEA data for 2015, nuclear power occupies here 3.5% of the total energy balance of the country. Currently, there are 22 operating reactors in the country, the total capacity of which reaches 6.2 GW. Five more nuclear reactors are at the construction stage; their total projected capacity is 3.2 GW. According to the nuclear energy program of India, by 2020, 14,600 MW of nuclear capacity will go into operation. The main goal of the program is that by 2050, the share of generation from nuclear power plants would account for 25% of all electricity.

D. Basu and V. W. Miroshnik, *The Political Economy of Nuclear Energy*,
https://doi.org/10.1007/978-3-030-27029-2_13

In 1924 SN Bose had analyzed the behavior of atoms under extreme temperature which is well known as Bose-Einstein Thermo Dynamics. By 1939, Meghnad Saha, Professor of Physics at the University of Calcutta, recognized the significance of the discovery of nuclear fission and started to conduct various experiments on nuclear physics. In 1940, the Sir Dorabji Tata Trust sanctioned funds for installing a cyclotron at the University of Calcutta, which was enhanced in 1973 onwards.

The peaceful atom is inextricably linked with nuclear weapons. In India, several events served as the motivation to master nuclear weapons. First, it is a territorial issue. As a result of claims by Beijing between India and China, two serious armed conflicts occurred. The first was in 1962 near Dhola and Khinzeman. The massive invasion of Chinese troops into India and bloodshed made the world community anxious. The USSR did not support its bloc ally in the conflict with India, called for a cease-fire and start negotiations. In addition, at this time, the United States and the United Kingdom have already begun to supply weapons to India. So, on November 21, a radio message about a unilateral cease-fire was heard on Beijing radio. In the Fall of 1967, the Cholsky incident occurred, in which 62 Indian soldiers were killed. However, it all ended again with the tactical victory of India. Thus, one of the motivations for India to acquire nuclear weapons is the potential threat from nuclear China.

Secondly, the division of British India into modern India and Pakistan on August 15, 1947, can be considered as another critical motivational factor. This event provoked tough clashes between Hindus and Muslims, as well as huge flows of refugees. About 500 thousand people died during clashes on religious grounds; millions were forced to leave their homeland. Frontier skirmishes systematically continued for more than half a century. Therefore, the creation of nuclear weapons was necessary for an advantage over the enemy.

Buddha Smiles While Is Bombing

The first works of Indian scientists in the field of nuclear science and technology were undertaken in March 1944; then for the first time, they proposed the creation of a nuclear research institute. In 1945, the leadership of India opened the Atomic Energy Research Committee to organize training in universities.

The purchase of uranium was made on the international market, as the Indian deposits of uranium are of too poor quality. The Indian Nuclear Testing Program was named the Smiling Buddha. On May 18, 1974, a nuclear device of its own design was blown up in the desert of the Indian state of Rajasthan. Preparations for nuclear testing, India conducted in the framework of the strictest secrecy. For example, the Prime Minister of Pakistan Zulfikar Ali Bhutto found out about what happened only from the newspapers. Tests have become a demonstration that India now also has nuclear weapons. In 1998, the Indians conducted a new series of tests called Testing to demonstrate that India now also has nuclear weapons. In 1998, the Indians conducted a new series of tests called "Shakti," and only after that India officially declared itself a state with nuclear weapons.

India on the World Nuclear Stage

As for the Treaty on the Non-Proliferation of Nuclear Weapons, which India does not want to sign, it entered into force in 1970. According to the document, the countries possessing nuclear weapons are those who produced and blew it up before January 1, 1967, and this is the USSR (1949), the USA (1945), the United Kingdom (1952), France (1960) and China (1964) In this case, if India signed the Treaty, it would have to terminate its nuclear program. According to the Indian side, the Treaty divides countries into privileged, having nuclear weapons, and into all the rest. Due to the fact that the country did not sign the document, access to civilian nuclear technology has long been organic for it.

In 1975, in response to Indian nuclear testing, the Nuclear Suppliers Group was established to limit the risk of proliferation of nuclear weapons by controlling the export of critical materials, equipment, and technology. In 1998, the United States imposed punitive sanctions against India for conducting nuclear tests. But relations between New Delhi and Washington developed dynamically, and in 2008 the countries signed an agreement on cooperation in the field of civilian nuclear technology, after which the

NSG allowed India to cooperate with other countries. On May 12, 2016, India applied to join the Group. The Americans supported the decision of the Indian authorities and the Swiss, and New Zealand, Ireland, Turkey, South Africa, and Austria spoke out. Today, surprisingly, it is China that is the main negotiator with India on membership in the Nuclear Suppliers Group.

Indian Reactors

The construction of the first research nuclear reactor "Apsara" began in 1955 with the support of British experts. Canadians agreed to sell India a 40 MW reactor, but with the condition that it will be used for peaceful purposes. The United States supplied 21 tons of heavy water. As a result, the reactor was named "Cirus"—Canada-India Reactor, United States. Later, the Indian side took all measures to ensure that no one could control the work on the reactor. It possessed considerable research power and was necessary for the possibility of creating weapons, where later plutonium was produced for testing in 1974.

The construction of the first nuclear reactor, used for civilian purposes, began in 1964. That was the Tarapur nuclear power plant located in the Indian state of Maharashtra. In 1969, two BWR-type units were built here, each of them was 150 MW. These nuclear reactors were the first not only in India but also in the whole of Asia. Initially, the designed power of the reactors reached 210 MW, but when the power units were launched, the specialists had to lower the capacity to the current 160 MW. The boiling pressurized water reactors of the two first power units are similar to the first rector of the Fukushima nuclear power plant, where in 2011 an accident occurred. After checking in the same year, Indian experts said that the reactors are very outdated and it's time for them to be decommissioned.

Later, the station began construction of two more units of the PHWR type, with a capacity of 540 MW each. Reactors were built in record time: in seven months. The third unit was launched in 2006, the fourth—in 2005. Currently, the total capacity of the Tarapur NPP is 1400 MW. All spent nuclear fuel (SNF) from this nuclear power plant is stored in storage facilities. India and the United States are negotiating the reprocessing of

SNF. If agreements are reached, India will have the opportunity to reduce nuclear waste and re-use used fuel. So the country will be able to save on the purchase of new fuel and accelerate the development of nuclear energy in general.

India has few natural reserves of uranium, which has long been a negative factor for the development of a peaceful atom. In this regard, the country is following its own unique ways of developing its nuclear energy. There is an alternative: the development of the nuclear industry with the help of thorium, a heavy radioactive metal. The stock of this substance in the country is much more. India is the second most important possessor of thorium in the world after Australia. So, India has been developing the nuclear fuel cycle, using its own reserves of thorium.

In the 1950s–1970s, for example, in the USA, Great Britain, India, Germany, and the USSR, various experiments were conducted with thorium and thorium-uranium reactors. However, it is not possible to move forward significantly by applying this technology. Thorium-232 itself does not support a nuclear chain reaction and cannot be material for an atomic bomb, however, if thorium is irradiated with neutrons, its atoms, capturing these neutrons, decay with the release of a significant amount of energy.

India suspected by the world community in such activities, which are not approved by the Treaty on the Non-Proliferation of Nuclear Weapons (NPT). And since New Delhi did not sign the Treaty, all the efforts of the NPT countries were directed not toward direct accusations of India in such actions, but more—in creating prerequisites for preventing India's free access to fissile materials and technologies for their production and processing. At the same time, after the nuclear tests, New Delhi had problems with the supply of uranium from the United States, which was resolved only in 1982, when India and France signed a long-term agreement.

On personal experience, the Indian authorities realized that it was unreliable to deal with imported nuclear fuel. The way out is to use natural, unenriched uranium. In this regard, the choice fell on the main reactor of the Indian nuclear program CANDU. Reactors of other types use only enriched nuclear fuel. Also, unlike most water-cooled reactors, CANDU is a channel reactor, which makes it possible to replace used fuel with fresh fuel during its operation, and the primary coolant can be either heavy or ordinary water. Thus, in 1963, the construction of the Rajasthan NPP began in the state of Rajasthan on the Chambal river.

The station was commissioned in 1972, and it was the first to use this type of reactor. The NPP includes two 220 MW CANDU type reactors, commissioned in 1973 and 1981, respectively. Canadian firms produced most of the equipment for them. However, after a test bomb of a nuclear bomb in 1973, relations between Canada and India collapsed, and the rest were created without Canadian participation. And yet, despite the "assembly of Canada," power units worked only by 40%, constant breakdowns accompanied the operation of the station. On the one hand, the reason for this was the termination of cooperation with Canadians. On the other hand, poor connection to the power grid station.

In 1990, within the framework of India's nuclear program, two more PHWR reactors with a capacity of 220 MW each began, which cost the country $570 million. The operation of the units started in 2000. Two and three years later, two more reactors of the same type and the same capacity were built, and in 2008 they began their operation. Thus, today six units with a total capacity of 1080 MW are operating at the Rajasthan NPP, and two more are being built: the sixth with a capacity of 220 MW and the seventh with 700 MW. To finish their construction is planned in the current year.

Another remarkable station is the Kudankulam Nuclear Power Plant, located in the south of the Indian state of Tamil Nadu. This station with two reactors of the VVER-1000 RU type with a total capacity of 2000 MW is the main object of cooperation between Russia and India in the field of nuclear energy.

Back in November 1988, Rajiv Gandhi and Gorbachev signed the Interstate Agreement, under which the construction of nuclear power plants was to begin. The construction of the Kudankulam NPP began ten years later. The process was delayed due to the political and economic situation in Russia and because of the position of the United States, who believed that the Russian–Indian agreement did not meet the requirements of the Nuclear Suppliers Group. In 2002, the Indian Atomic Energy Corporation entered into a contract with Atomstroyexport to develop working documentation and other things.

With the launch of this nuclear power plant, everything somehow did not evolve: first, the change of power in Russia, then the position of the United States, then the protest movements. The first connection to the network of NPP Kudankulam successfully took place only on October 22, 2013, in accordance with the test program at a minimum power of 160 MW. The first unit was commissioned entirely on August 10.

Also, according to the US energy engineering magazine Power Engineering, the Kudankulam NPP was named one of the best projects in the Nuclear Power Industry nomination. As noted by the publication, this project is the embodiment of the newest technologies, allowing to create the most environmentally friendly and efficient energy sources.

The second unit of the Indian nuclear power plant has already been built. The regulator of India gave the go-ahead, and its connection to the electricity grid took place on August 29, 2016. Initially, the cost of building two units of the Kudankulam NPP was estimated at two billion dollars each; however, after the protests, the cost of building the station increased by another half a billion dollars.

Apparently, the Kudankulam NPP will not be limited to only two units. So, in 2014, Russia and India signed an agreement on the construction of the second stage of nuclear power plants, which will include two more units with a capacity of 1000 MW each. And on August 16, 2016, it was reported that Rosatom would build two units of the Kudankulam NPP at the cost of approximately six billion dollars. For the third and fourth blocks, this year, soil preparation was started. Construction began in 2017. Kudankulam-3 NPP and Kudankulam-4 NPP are planning to be commissioned in 2022 and 2023, respectively. True, in comparison with the first stage of the station, there will be a significant difference. If the localization of the equipment of the first two energy blocks was zero, then now India has the necessary power.

Today, despite India's desire to develop its nuclear industry, coal, oil, and gas continue to occupy a dominant position in the country's energy sector. From 2012 to 2017, India's dependence on oil imports will increase from 76 to 80%, natural gas from 19 to 28%, coal from 19.8 to 22.1%. However, in New Delhi, it is planned to increase nuclear generation by 14,600 MW by 2020, so that by 2050 the share of the peaceful atom in the country's total energy balance is at least 25%. That requires significant investments, primarily foreign ones, since NPPs are expensive, as well as establishing connections in the global nuclear arena. Therefore, joining the NSG is a correct step. And given the unique experience of India, we can confidently say that the nuclear ambitions of this South Asian country have not shaken at all.

March 2019 United States to Build Six Nuclear Power Plants in India

American companies will build six nuclear power plants in India. During the strategic dialogue, the two countries considered issues of cooperation on a peaceful atom and confirmed plans to build six American nuclear power plants on Indian territory.

Earlier it was reported that the Indian government issued a permit to build 12 nuclear reactors in the country and approved the appropriate financial support.

Russia and India signed a general framework agreement for the construction of the third stage (power units NN5-6) of the Kudankulam NPP in India. The current stage of development of the Indian atomic project, in fact, started in September 2008. It was then that the Nuclear Suppliers Group (NSG) removed restrictions on the supply to India of goods from the Group's control list. If you remove official verbal decorations, from this point on India, received the right to import into its territory everything necessary for the development of nuclear energy, including nuclear fuel and uranium enriched to energy levels.

The NSG was established in 1975—right after India conducted a nuclear test a year earlier. At that time, the NPT was already in force, and a group of countries—nuclear suppliers, decided to move to practical measures for its implementation. The NSG rules only allow cooperation in the nuclear field if the recipient country has put all its nuclear activities under comprehensive IAEA safeguards. This rule does not apply to contracts and contracts concluded before April 4, 1992, and for newly joined states prior to their accession to the Group.

Russia was one of the founders of the NSG, all new states gradually joined the group, and now there are 48 members in the NSG. In 1975,

India came under the "blow" of the NSG, and nuclear export to it was banned. Since India has minimal uranium deposits in India, after all, things in its nuclear power industry went unhappily, sometimes it came to forced shutdowns of reactors.

In 2008 the NSG, despite the fact that India did not accede to the NPT, removed its restrictions. India agrees to separate its military and civilian nuclear facilities and put the latter under the control of the IAEA.

The price of such obedience was also announced—the United States pledged to expand full-scale civil nuclear cooperation with India. The nuclear agreement with the United States was signed almost immediately after the Joint Statement, a week later, the NSG responded by lifting the restrictions.

In order to save its nuclear power industry, India was forced to agree to sign contracts with Westinghouse and AREVA to build an APR-1000 and EPR based nuclear power plant on its territory. India and its leading elite, of course, are not something homogeneous, having a single view of the prospects of their country. At that time, in the Indian nuclear industry, the prevailing view was that it was necessary to import the maximum possible atomic energy technologies to be able to evaluate each of them and choose the best one. But, despite this trend, there was a group of specialists who completely disapproved such an approach. They suggested concentrating on developing their own heavy-water reactor (PHWR) projects, while at the same time working on fast reactors and thorium reactors. But this group did not intend to abandon imported reactors either, but these specialists considered the Russian VVER-1000 to be the ideal "kit" for their own reactors.

So, in 2008, India signed an agreement on comprehensive cooperation in the nuclear power industry with the United States, and almost instantly, with unprecedented speed, already in December 2015, informed circles reported that India was about to be ready to sign an agreement on the construction of APR-1000 with the company Westinghouse Electric.

In January 2015, Barack Obama arrived in New Delhi and among other things, signed a framework agreement with a private, independent of the US government, Westinghouse. The extremely truthful edition of Bloomberg came up with a rump headline: "India has signed a $182 billion contract with the United States to build nuclear power plants."

One hundred eighty-two billion is 18 APR-1000 reactors, and the word "contract" implies the mutual obligations of the parties.

The fact that the Indian Nuclear Energy Corporation and Westinghouse Electric can start the design and planning of construction sites in India for APR-1000 reactors in the amount of 6 (six) pieces. Negotiations with France and its AREVA after the 2008 transaction proceeded with the same lightning speed. In February 2009, a framework agreement was signed on the construction of two EPR-1600 reactors in India, while the French showed confidence that this was only the beginning, that the number of units would increase to six.

In 2011, a catastrophe occurred at the Fukushima nuclear power plant, after which India asked the French to provide the expertise of the IAEA that nothing like this could happen with their units. Most of the territory of India is located in the seismic zone of medium or high risk.

Besides Westinghouse and AREVA there was another consortium that intended to help India with the development of atomic energy—GE/Hitachi. In 2007, these companies created a joint venture with the abbreviation GEH with the goal of building nuclear reactors around the world. Already in 2015, almost instantly, GEH received a certificate from the US Nuclear Regulatory Commission for the design of an improved boiling ESBWR.

The main goal of India is to reach the total NPP capacity of 63 GW by 2032. As of March 2017, 22 industrial nuclear reactors with a total capacity of 6.2 GW were operating in India. On May 17, 2017, the Indian Cabinet approved the construction of 10 heavy water reactors of the Indian project PHWR-700.

India as the Leader of Thorium Energy

In October, the world's first commercial nuclear reactor will be launched in southern India, producing, along with electricity, a new nuclear fuel from thorium raw materials.

Kalpakkam is located in the south Indian state of Tamil Nadu; now it is called the atomic capital of India. It was there that in 1971 the Reactor Research Center was founded since 1985 it has been called the

Indira Gandhi Center for Atomic Research (IGCAR). In the same place, in Kalpakkam, Madras NPP operates, consisting of two power units with an electrical capacity of 220 MW each. They were commissioned in 1984 and 1986. These are the first blocks designed and built by Indians entirely independently.

In 2004, the construction of a prototype fast breeder reactor (PFBR) with a capacity of 500 MW, developed by IGCAR, began at Madras NPP. In February 2017, construction was announced and planned to put the PFBR into operation in October 2017. Department of Atomic Energy of India would build two more PFBRs with a capacity of 600 MW each in Kalpakkam.

Payback for "Smiling Buddha"

The Indian atomic program was started back in the 1950s but immediately ran into a number of problems. Some of them, such as lack of financial resources, lack of highly qualified personnel, dependence on foreign technologies, took decades to solve. However, two problems, apparently, will remain with India forever. First of all, this lack of sufficient reserves of natural uranium. As a result, India is dependent on foreign suppliers of nuclear fuel.

India's second problem, inherited from the time of Indira Gandhi, is its categorical non-participation in the NPT. This treaty was approved by the UN General Assembly in 1968 and has now been signed by 191 countries. Ukraine, we recall, joined it in 1994 in exchange for "guarantees" under the Budapest Memorandum.

The Treaty imposes very strict requirements on the control of the circulation of fissile materials to prevent the expansion of the circle of nuclear-weapon States. Such countries, according to the treaty, are those that made nuclear explosions before January 1, 1967, that is, the USSR (and its successor Russia), the USA, Great Britain, France, and China. Israel, India, and Pakistan refused to sign the treaty. North Korea withdrew its signature. These are the states that acquired nuclear weapons later than the "legitimate" five nuclear powers.

India made its first nuclear test on May 18, 1974. Then, in the Thar desert in Rajasthan, an 8-kiloton nuclear charge was detonated. After this test, called "Smiling Buddha," it became clear that neither the United States nor even the Soviet Union would give India any technology for the separation of uranium isotopes.

The only isotope with which to start the nuclear cycle is uranium-235. Its share in natural uranium is only 0.72%, and the remaining 99.28% is uranium-238, which is not nuclear fuel. Conventional nuclear technology is that natural uranium is enriched (that is, it increases the proportion of uranium-235) to produce energy at nuclear power plants, and the remaining "lean" mass (mainly uranium-238) is thrown into waste.

Workaround Through CANDU

Nevertheless, India has found ways to circumvent both of these problems. As for the separation of isotopes, it is not a prerequisite for the development of nuclear energy. Back in the early 1960s, a PHWR was created in Canada—a heavy-water pressure reactor called CANDU (Canada Deuterium Uranium). It uses ordinary natural uranium as a fuel. India obtained this type of reactor in the framework of cooperation with Canada in the early 1970s, but after the "Smiling Buddha" cooperation was curtailed on the Canadian initiative, and India continued to operate and further modify the original CANDU reactor design itself. The power units of the Madras NPP were the first of this series.

Currently, in India, there are 22 power units at seven NPPs with a total capacity of 6780 MW or 6.78 GW. Of these, 18 blocks—type PHWR. The construction of four more units of the same type of 700 MW is nearing completion. And in May 2017, India approved the construction of 10 more PHWR reactors of 700 MW each. Thus, after the implementation of all these projects, including the launch of the PFBR in Kalpakkam, the total capacity of Indian nuclear power plants will increase by 2.5 times compared to the current one.

Fuel Underfoot

Indian thorium reserves are located in the very south of the country in the form of alluvial sands of a very dark (almost black) color with a characteristic metallic luster, much heavier in weight than ordinary sand. They are formed by the erosion of volcanic rocks in coastal waters.

In India, not only the extraction of thorium sand on the coast and the extraction of thorium from it but also the manufacture and processing (reprocessing) of thorium assemblies at a radiochemical plant in Tarapur, have long been established. It is planned to use these assemblies in new PHWR reactors to equalize the energy release already in the first load, which will allow outputting such units to their rated capacity at the initial stages of the start-up program.

However, the results of using different mixtures in the PFBR breeder are promising to be especially interesting. Plutonium-239 should be obtained from uranium-238 in the zone of reproduction, and 233 from uranium-233 and both final products can be used as nuclear fuel. The newspaper emphasizes that China is now trying to master the same technology, but it is "more than a decade behind India." India, having created the PFBR, was "an innovator of nuclear technology that could potentially become the largest source of energy in the country."

In April 1955, the Canadian government offered to assist in building an NRX-type reactor for India. CIRUS was completed in early 1960. Since 1974 when India exploded its first nuclear bomb, all collaborations with the West came to an end and USSR became the sole supplier of all nuclear facilities to India. India received natural uranium, enrichment plant and Fast Breeder Reactor from the USSR in 1977.

India is a leader on thorium-based research, and no other country has done as much neutron physics work on thorium.

Russia wanted to supply on-shore nuclear power plants with added safeguards that the plants cannot be used to produce any nuclear weapons. Without the nuclear deal, India would be able to maintain its nuclear plants by using reprocessed plutonium as fuel and using its own uranium in the conventional plants. Russia can still supply onshore and off-shore nuclear power plants, whatever India needs at a much lower price. It will continue to get offshore nuclear plants from Russia. In that case, it will be at liberty to test further nuclear weapons in the future. That is precisely what President Putin has suggested, but India was not interested signing the deal with the US Westinghouse company. Now when the Westinghouse went bankrupt the future of Indian nuclear industry remains unclear.

14

China as Nuclear Power

China's nuclear strategy is based on the development of nuclear energy, which so far only takes 2% in the production of electricity in China. The main source of energy is coal, which accounts for 80% of the energy balance. The new Chinese authorities understand the unpromising ecological nature of the "black energy" and are actively developing nuclear power. By the beginning of 2018, 38 nuclear power units were in operation in China, which is about 9% of all capacities in the world.

France has the record of supplying nuclear materials and training to China who exploded nuclear devices in 1964 without having a nuclear reactor then. France made a nuclear reactor for China in 1994 and helped China to develop nuclear weapons. By 2012, 15 power units have been built and put into operation in China by six operating nuclear power plants. By 2015, the total capacity of Chinese NPPs would have reached 40 million kW, and by 2020, 70 million kW.

China Nuclear Energy Association said that nuclear energy generation in 2017 increased to 247.5 TWh gross, which is 4% of total production. Technology came from France, Canada, and Russia. The latest technology is coming from the United States (via Westinghouse), owned by Japan's Toshiba) and France. China also has a policy to export nuclear technology.

D. Basu and V. W. Miroshnik, *The Political Economy of Nuclear Energy*,
https://doi.org/10.1007/978-3-030-27029-2_14

In 2015 the Hualong One reactor became the main export product to Pakistan.

In May 2014, the China Atomic Energy Authority (CAEA) agreed with ROSATOM to cooperate in the construction of floating nuclear cogeneration plants for China offshore islands. These planned to built in China but be based on Russian technology, and possibly using Russian KLT-40S reactors and Russia is anticipated to provide fuel for them.

A weak spot is the inadequate quantity of natural uranium deposits. In 2011, China produced between 1350 and 1500 tons of uranium; this is the 9th among uranium-producing countries. Partially, the natural defect is compensated by active cooperation with Kazakhstan and other countries in the joint production of uranium and its imports, as well as the introduction of new Chinese deposits. Astana plans to increase in 2014 the export of uranium fuel pellets to China from 2 to 200 tons.

China actively replicates and adapts advanced French and American technologies for the production of reactors for their own needs. By 2020, the Chinese intend to establish the export of Chinese reactors. In 2000 and 2011, the PRC, based on its technologies (nuclear reactors Chamsha-1 and Chamsha-2), has already implemented the construction of two power units in Pakistan (near Karachi). However, the security regimes of these stations, according to IAEA experts, require additional refinement and verification (inspection). Fukushima's and Chernobyl's Syndrome increase these fears.

Russian expert Andrei Gubin notes that the current export potential of China in the field of nuclear technology "*can be assessed as low - the proposed models are obsolete and do not meet modern safety standards*" (https://nsarchive2.gwu.edu/).

The scientist believes that it will take China about 15 years from 2018 to reach the production of the newest third-generation reactors and to make real competition for France and Russia in the markets of nuclear reactor technologies. Meanwhile, leaders should be afraid of unlicensed copying by China.

At the same time, the Chinese are strong in the skillful and flexible use of their financial capabilities. A typical example is the successful investment project of the state company Chinese General Nuclear Power Group (CGNPG). After long negotiations, the company's management convinced the British authorities of the need to participate in the construction of the British nuclear power plant in Somerset (U.K.) as the main investor of the project at a total cost of $21.5 billion. According to the contract, the investor gets access to the newest nuclear technologies.

According to a similar model, since 2009, there have been talks between the Chinese and the French (EDF). We are talking about the joint creation of two power reactors on special technologies and Chinese money.

For Russia, the Chinese nuclear market is of particular importance. Experts believe that China is still half as far behind Russia in the development of peaceful atom, which gives it additional opportunities in terms of the quality of technologies, services, and a high level of security. Shortly, however, it is possible that the leaders of this market will have to face the Chinese competition.

There was a Russian–Chinese agreement of 1992 on the supply of two reactors and the construction of the Tianwan NPP near Shanghai. A 3.2 billion contract was implemented, and in 2007, two power units of nuclear power plants were constructed, and construction started for the 5th and 8th power units.

Chinese Nuclear Weapons

The Chinese military atom is not so clear and transparent. On the one hand, China is a member of the official "nuclear five," including the USA, Russia, England, and France. After conducting a series (from 1964 to 1996) of nuclear tests from 45 ground and underground explosions, in 1992 the People's Republic of China ratified the Treaty on the Non-Proliferation of Nuclear Weapons (NPT), undertaking the obligation not to use the first nuclear weapons.

On the other hand, in the conditions of the country's withdrawal to the global status of a superpower, China faces an objective need for a sharp qualitative strengthening of its strategic nuclear status. While it is much inferior to traditional leaders—the United states and Russia, many Chinese experts call this lag the weakest point "in the strategy of elevation," pointing to the inevitability of a "future confrontation and even conflict with the United States." Some radical scientists believe that by 2020, China will not only reach a comparable level of strategic arms with leading powers but also provide “restraining strategic influence on the United States,” replacing Russia in this capacity.

How Many Nuclear Missiles Does China Have?

The nuclear program of China was first implemented with the help of the USSR. During the 1950s and 1960s, 10,000 Soviet nuclear specialists visited China. About 11 thousand Chinese specialists and a thousand scientists of this industry were trained in the Soviet Union. On April 7, 1956, the first Soviet–Chinese agreement on cooperation in the nuclear industry was signed. October 15, 1957, signed a second agreement, which directly provided for the transfer to China of nuclear weapons technology (the USSR signed the NPT in 1968). After the complication of Soviet-Chinese relations in the 1960s, aid from the USSR was curtailed, Soviet specialists withdrew, but this could no longer halt the progress of China’s nuclear project.

The project was crowned with success—the first test was held on October 16, 1964, with the help of France. On June 17, 1967, China successfully carried out a thermonuclear analysis, again supported by France. That is the shortest period between the first nuclear and the first thermonuclear test in history: it took China only 32 months to make a nuclear fusion based on nuclear charge, although at that time China had no nuclear power plant or the nuclear fuel cycle. France supplied all the materials.

Some sources mentioned the Chinese test of the neutron bomb on September 29, 1988. General Zhang Aiping, head of the Science and Technology Commission of the Ministry of Defense (from 1982 to 1988, the PRC’s defense minister) allegedly told the US delegation in 1980 that China needed a neutron bomb to defend itself from the Soviet Union (https://nsarchive2.gwu.edu/). The last nuclear test was conducted by China on July 29, 1996, after which it signed in September 1996 a nuclear test ban treaty. In total, China conducted 45 tests, 22 of them underground.

Soviet specialists helped China and with the development of missile technology. The USSR gave the Chinese two P2 missiles (based on V-2) and a full set of documentation, and the documentation for the R-12 missile, which allowed the PRC to create a medium-range missile Dongfeng-1, adopted in 1970 for service. The basis of China's strategic aviation is the H-6 bomber, based on the Tu-16, originally from the USSR.

In the 1990s, China declared its doctrine of non-use of nuclear weapons (N.W.) first under any variant of a non-nuclear attack, as well as on the complete abandonment of threats to its use in countries that do not have nuclear weapons.

Along with these statements, hundreds of spies from China are scouring the United States in search of nuclear technology. Back in the 1980s, Chinese scientists, while on an official visit to American nuclear laboratories, managed to steal some of the information related to the creation of the neutron bomb, as well as the W-76 warhead drawings for the Trident-2 missile. Spies from China often act with the help of US citizens of Chinese origin, such as Guo Bao Ming, a scientist from the Livermore Laboratory, where he was fired in 1981 for espionage (stole data on the W-70 neutron warhead). The most surprising is that when such Chinese spies and agents are caught red-handed (from 2012 to 2016, about 100 people were arrested in the United States for espionage in favor of China), they receive punishments either not strict or justify them. Especially infamous was the case of the theft of technical documentation for the most advanced thermonuclear warhead W-88, which allowed China to save billions of Dollars.

The US intelligence services paid attention to the physicist from Los Alamos Wen Ho Lee, after his telephone call to the Guo as mentioned earlier Bao Ming. At the beginning of the conversation, Lee directly called himself a nuclear weapons designer. Investigation revealed that on the theft of the W-88 warhead documentation, Lee in 1994–1995 sent secret files to the computer network of another laboratory. Li was arrested on December 10, 1999, on suspicion of spying for China. He spent nine months in solitary confinement. But the accusation of espionage was refuted; the scientist was accused of "only" transferring sensitive information from a protected system in Los Alamos to his laptop.

At the same time, China keeps its nuclear secrets perfectly: the Chinese leadership has never even voiced the approximate size of its nuclear arsenal anywhere. That is the basic principle of China's nuclear policy: do not call the number of nuclear charges and ballistic missiles. This simple principle is of particular importance in assessing China's nuclear arsenal. In other respects, the above-described history of China's nuclear weapons is no worse than that of other countries, moderately dramatic and quite successful.

China's Strategic Nuclear Forces

After China tested a solid-propellant mobile intercontinental ballistic missile (ICBM) with a detachable warhead capable of "reaching" anywhere in the United States, there was a rumor in the world that now Beijing has a powerful nuclear deterrent and the Americans will have to reckon further with the new status of the Asian dragon.

However, if we discard the propaganda husks and analyze the real state of affairs in the Chinese strategic nuclear forces, a completely different picture opens up: China has simply "eaten up" the remnants of Soviet technologies kindly provided in the 1950s and 1960s, as well as those sold by Russian specialists in the era of the chaos of the 1990s. From scratch, the USSR built whole industries in China—rocketry and nuclear fusion—with hundreds of factories and scientific centers, trained tens of thousands of Chinese specialists and transferred all the technical documentation to all of its projects.

United States–China: Comparison of Nuclear Forces

The reality is that with a real nuclear war, against the United States, China will not last even a few hours. The modern Chinese military industry was born in the 1950s with the help of the USSR. USSR helped China to establish a modern military-industrial complex capable of producing the entire range of military products needed by the Chinese National Liberation Army. The figures are impressive: the Soviet Union in China has built 763 full-fledged plants with the entire infrastructure and the most modern equipment, 97 scientific and technological centers, 11 test sites, including four underground ones.

More than 120,000 Chinese students studied free of charge in Soviet technical universities in military subjects, and about 6000 domestic scientists, 85,000 technologists, and other professional specialists visited China on long-term business trips. The enterprises built at that time, for example, aviation complexes in Shenyang, Harbin, Xi'an and Chengdu, a tank factory in Baotou (Inner Mongolia, the so-called factory No. 617), a complex of enterprises for the production of small arms and artillery weapons in the north-east of the country and many others are the basis of the Chinese military-industrial complex.

Russia–China: Comparison of Nuclear Forces

The USSR transferred China licenses for the production of a full range of weapons and military equipment—from aircraft to communications equipment and engineering equipment. The strategic armament also remained on the sidelines: before the Soviet-Chinese rupture, China managed to get a huge array of documentation and equipment needed to create a complete production cycle of nuclear weapons.

In 1951, a secret agreement was signed between the USSR and China on providing scientific and technical assistance to the Chinese in the field of nuclear research in exchange for deliveries of uranium ore, within the framework of which the USSR transferred to China uranium enrichment technologies, centrifuge construction and other stages of the production process. By the way, the centrifuges of the last generation of Soviet manufacture were finally brought from Moscow and given along with all the necessary documentation, because Chinese specialists could not master their production in their homeland.

The Chinese general in charge of the nuclear project, Hung Zi in a letter to Lavrenty Pavlovich Beria wrote: "*The process of mastering the production of gas centrifuges cannot be adjusted. I ask again to send to Hayon a group of specialists for debugging equipment and training our engineers. Unfortunately, it will be necessary to shift the start-up period to the industrial operation of the plant for the fifth time. I hope that this time, technologists will still be able to transfer all the subtleties of technology* fully" (http://www.authorstream.com/Presentation/Reva-20059-InzhechikBaksan07).

In 1957, the USSR concluded another agreement—on the transfer of the Celestial Soviet missile technology with a full cycle of training of local specialists by the forces of Soviet universities. Within the framework of this agreement, the Beijing Institute of Physics and Atomic Energy was established, where nuclear research began, and in Lanzhou—the construction of a gas diffusion plant for uranium enrichment.

A turning point for the Chinese was the launch at the plant No. 601 in the capital of China of the experimental heavy water nuclear reactor supplied by the Soviet Union. In 1958, under cover of the strictest secrecy, the Soviet government opened a nuclear test site near Lake Lobnor in the Xinjiang Uygur Autonomous Region, where the Chinese strategic forces are still carrying out all their tests.

Since the death of Stalin, China became hostile to the USSR because of the Soviet support for India and the USSR, in turn, refused to provide any large-scale protection to China, which immediately affected the course of the entire nuclear program of China. While the USSR has already begun mass production of mobile missile kits with an active system for installing false targets and missiles with a range of up to 15 thousand kilometers, on October 27, 1966, the first Chinese strategic ballistic missile Dongfeng delivered the 12-kiloton uranium warhead, with the materials supplied by France, at a distance of 894 km, developed on the basis of the Soviet R-5 M model in 1956. The Chinese defense industry, for the first time, managed to create a full-fledged nuclear missile weapon.

Since then, China's nuclear missile forces have not changed in quality. Vasily Kashin of the Institute of Far Eastern Studies of the Russian Academy of Sciences, described their condition as follows: "*Until now, China's strategic nuclear forces consisted of fairly primitive and bulky intercontinental ballistic missiles*" (http://russiancouncil.ru/en/inner/?id_4=7153#top-content).

Nuclear Weapon Policy of China

China has repeatedly stated that the Chinese nuclear forces are small, and besides, it will never apply them first. Usually, the official nuclear doctrine of the People's Republic of China is perceived by experts as a particular propaganda document, since the strategic nuclear forces of the People's Republic of China have real value only with a preemptive strike. There is a significant amount of assessments of the current state of Beijing's nuclear potential, which sometimes differ by order of magnitude and offer highly contradictory pictures.

China is considered itself the new center of power of the modern multipolar world. At the same time, as the United States–China rivalry grows the danger of destabilizing the situation both in the Asia-Pacific region and in the world increases. The economic growth of China is accompanied by an objective strengthening of geopolitical ambitions and active construction of the armed forces. Under the new conditions for China, the only way to contain the United States and its allies, and in the future, perhaps, maintain a global status, is strategic nuclear forces.

There is a significant amount of assessments of the current state of China's nuclear potential, which sometimes differ by order of magnitude and offer highly contradictory pictures. Formally, the strategic nuclear forces of the People's Republic of China or the "Second Artillery Corps," which is now being transformed into a separate type of armed forces "Rocket Forces," are nuclear dyad-ballistic land-based missiles and ballistic missiles of submarines. The Air Force is capable of performing tactical tasks with the use of nuclear weapons, but strategic capabilities have not been reliably confirmed. The total potential is estimated at 400 nuclear munitions, of which 260 are formally on strategic carriers. Meanwhile, there are different opinions on this matter.

For example, one opinion is that China as of 2010 had only 240 nuclear warheads, of which only 175 were on duty. Another opinion is that China possesses more than 3500 units of nuclear weapons, with 200 new generation warheads produced per year. For each of the launchers, there are up to five missiles, which allegedly indicate the intention to conceal the actual size of the arsenal, which is usually measured by the number of carriers, and the readiness to strike a nuclear strike in several waves.

It seems more realistic that the nuclear potential of China does not exceed 300 munitions on strategic carriers, including free-fall bombs of 15–40 kiloton and three megaton, warheads of missiles with a charge of 3–5 megaton and more modern 200–300 kiloton. 150 ammunition can be deployed on medium- and shorter-range ballistic missiles, and possibly in cruise missiles, while for a long time, China's strategic nuclear forces consisted of rather primitive and bulky liquid ICBMs that require several hours of preliminary training, since the warheads are stored separately and the rocket requires refueling. According to the scenarios described by the experts of the Monterey Institute of International Studies F. Sanders and C. Yuan 15 years ago, by 2010 China had to reach the level of minimal deterrence, possessing about 50 ICBMs, mainly mobile solid-fuel complexes "Dunfeng-31" with a range of 8000 km and carrying one warhead (immediately installed, and the missile is ready for launch within 10 minutes). Only about 18 old DF-5a liquid-propellant ICBMs with a range of more than 13,000 km are available.

According to the forecasts of American experts, by 2020, China can achieve the potential of so-called "doctrinal" or limited nuclear deterrence. On alert, there will be up to 200 ICBMs, both mine and on the automobile chassis. The basis will be the Dongfeng-31HA and Dunfing-41 complexes with a range of 11 and 14 thousand km respectively, and the latter can carry up to 10 warheads (both warheads and false targets). The main planned innovation is the early warning system on missile attack, which China does not currently possess.

The greatest uncertainty and maximum veil of secrecy remain around the submarine fleet of the PLA Navy, especially its atomic component. The first Chinese nuclear-powered ballistic missile (SSB) project 092 "Xia" was commissioned in 1987 and equipped with 12 missiles "Jiulan-1" ("Big Wave") with a range of up to 2500 km. Until recently, she was not on duty, always defending herself at the base of Jiangeshzhuang near Qingdao. Probably, its exits in the sea are quite rare.

In May 2008, the PLA Navy conducted tests in the Yellow Sea of a new Julian-2 submarine (SLBM) ballistic missile (marine version DF-31, range 7400 km) intended for placement on the new SSB 094 Jin (12 missiles) and subsequent ones. In the island of Hainan, a vast subterranean submarine base was built, entirely closed for tracking from space. According to data at the beginning of 2010, China may have had three Jin class boats, how many of them are loaded with missiles and are on alert—it is not known. Current estimates are two boats constantly on patrol. By 2020, the number of SSBNs in the Navy of the PLA, according to US data, can be brought to eight. According to some reports, China is developing a new-generation SSBN project 096, the first of which can be commissioned in 2020.

The PLA, in all likelihood, intends to strengthen the possibilities for regional deterrence, which can compensate for the time being insufficient potential for the defeat of the US mainland. In January 2014, China tested a hypersonic missile that could carry both conventional and nuclear warheads, but subsequent launches were unsuccessful, and there is no precise information on developments in this area.

In 2014, China confirmed the presence of a certain number of medium-range missiles DF-26C (range 3500 km), the so-called "killers of Guam," with JAH. On ground launchers, from 2007, 40 to 55 CJ-10 cruise missiles with a range of 1500 km are deployed, their total arsenal is estimated at 500 units, they can be launched from the H-6 K bomber, and in the future from surface ships and submarines.

A kind of deterrent can serve as an anti-ship ballistic missile DF-21D, capable of hitting a mobile surface maneuvering individual target at a distance of up to 1500 km. The rocket has been dubbed the "killer of aircraft carriers." Non-strategic range missiles are very important for the nuclear forces because of the presence of US military infrastructure in North and SouthEast Asia.

Major R & D is currently concentrated around increasing the capabilities of missiles to overcome any missile defense of a likely enemy. In this connection, maneuvering warheads, light, and heavy false targets are being developed for deployment on ICBMs and SSBNs with MIRVs; anti-satellite weapons kinetic and electronic impact. China's military-industrial complex is working to improve the accuracy of strategic and non-strategic combat systems, including the improvement of correction and control systems.

China–Pakistan: Nuclear Weapon Program

Pakistan is well on its way, to gather the third largest stockpile of nuclear weapons with the support of China. China is supplying nuclear weapons and aircraft, missiles to Pakistan continuously.

Pakistan built, with Chinese help, the 40–50 megawatt heavy-water Khushab plutonium production reactor. Pakistan also built an additional three heavy-water reactors. Pakistan built a plutonium reprocessing plant (23 April 2018, www.military.com).

Besides, Pakistan started a program boosted by A. Q. Khan to produce HEU by using gas centrifuge enriched uranium. A. Q. Khan, a metallurgist, working in the British-based Uranium Enrichment Company (URENCO Group) in its Dutch branch, returned to Pakistan in 1975 with stolen blueprints for various centrifuge designs and a broad array of business contacts from URENCO.

The Pakistani nuclear effort received considerable assistance from China, who in turn received all kinds of help from France. China supplied a broad range of missile and nuclear weapons-related assistance to Pakistan. According to the information from the IAEA, Pakistan currently has between 140 and 150 nuclear weapons under its control. However, Pakistan stockpiled around 300 kilograms of plutonium and about 4000 kilograms of weapons-grade HEU. Pakistan's HEU based warheads utilize an implosion design that requires between 15 and 20 kilograms of HEU, hence the current stockpile is enough for an additional 250 weapons. The plutonium-based warheads need about eight kilograms of plutonium; thus, the existing stockpile would be enough for 40 other warheads. Therefore, Pakistan has enough HEU and plutonium to produce an additional 230–290 warheads by the end of 2018. However, this number could be higher if Pakistan opts for smaller warheads, resulting from raising the Pakistani nuclear arsenal to 450 warheads. Pakistan is adding enough HEU and plutonium to produce 20 additional bombs a year, which would be the third largest arsenal of nuclear weapons. Pakistan never signed the 123 Treaty with the United States, which India was forced to sign.

Nuclear Weapons of China: Concerns for Japan and India

According to experts from the authoritative edition of Jane's, the ICBMs with MIRVR on mobile platforms, as well as the new Chinese SSBNs on patrol, are the most dangerous for the United States. Over the next five years, the Chinese strategic nuclear forces will seriously undermine the American deterrence strategy and "put across" on Japan and US military bases in Northeast Asia and SouthEast Asia as "unsinkable aircraft carriers." Until the end of 2015, the Chinese satellite group can number up to a hundred devices, which can significantly strengthen the architecture of C4ISR and increase the effectiveness of all armed forces in general.

It is noteworthy that only the United States declared its concern about the growth of the potential of China's strategic nuclear forces. Japan expresses concern about the growing military capabilities of the whole of China, without commenting on its nuclear weapons, being itself a state that does not possess it. Potential regional rivals, the ASEAN countries, realizing the senselessness of military competition, leave the nuclear factor behind the brackets.

Taking into account the full development of the PLA, it seems that China will pay considerable attention to improving the nuclear deterrence mechanism, which is represented by the Missile Forces. The current situation of the US–China regional confrontation in the APR can be transformed in any way from the division of spheres of responsibility with Washington to the formation of new global military-political blocs.

In the future, China, possessing perfect and sufficiently numerous strategic nuclear forces, can become one of the guarantors of international stability, as it is implicit in its permanent membership in the U.N. Security Council and the destabilization of the international situation in the event of a more radical foreign policy line.

Russia is currently guided by considerations of "strategic partnership" with China and does not intend to regard it as a potential rival. At the same time, China's non-alignment with the Russian-American START process, now defunct, creates a situation of strategic uncertainty about the direction of the Chinese nuclear forces.

Peaceful Atom in China

To date, China has a whole range of problems in the energy sector. First, the most important problem is related to the strong dependence of Chinese energy on coal-fired power plants, which is becoming an obstacle to further sustainable development. With coal energy, serious environmental problems and logistical difficulties are associated, including, the congestion of the transport system by coal transportation from the coal-mining regions to the east of the country. Also, China has been a major importer of coal for several years (in 2010, this figure reached 166 million tons). Opportunities of China to develop hydropower are limited by natural factors: a significant part of the hydro resources is located in the seismic zone. In turn, the potential of other renewable energy sources is still insufficient.

In these conditions, the development of nuclear power is considered by the PRC as one of the most promising directions in ensuring energy security. Even though the commissioning of the first nuclear power unit in China was carried out in 1993, today the country is implementing the world's most extensive program for the development of nuclear energy, involving leading European, American and Russian manufacturers of equipment for nuclear power plants.

The nuclear industry of China includes geological prospecting and mining of uranium, its conversion, and enrichment, the operation of nuclear reactors, the processing of spent fuel, the storage of radioactive waste, the production of specialized equipment and instruments, the construction and installation of new installations and radiation protection.

At present, the nuclear industry of China is developing in three main directions:

- Power generation at nuclear power plants;
- Nuclear fuel cycle;
- Research, design, and engineering activities.

According to the World Nuclear Association, to date, 14 reactors operate in China, another 27 are under construction, and another 50 reactors are planned to be commissioned by 2030. Chinese authorities expect to bring the share of nuclear power plants in electricity generation to 5% by 2020, increasing their capacity by four times.

In order to achieve these objectives, in 2005 the State Council of the People's Republic of China adopted the "Medium-term and long-term development of nuclear energy program (2005-2020)", which, provides for an increase in the installed capacity of nuclear power plants up to 42 GW. In accordance with this program, the development of China's national nuclear power industry is based on the following provisions:

- The main type of reactor to be PWR 1000 MW;
- Increase the share of own equipment, expand the independence of the Chinese industry in the design, construction, operation and preventive maintenance of its nuclear power plants;
- Reduce the construction time of the NPP;
- Ensure the competitiveness of nuclear power plants in comparison with coal-fired power plants;
- Develop new generation nuclear reactors (fast neutron reactors and high-temperature gas-cooled reactors).

Priority tasks were identified as follows:

- Reduction of harmful emissions into the atmosphere;
- Increasing the technological level of the energy industry;
- Creation of reactor equipment production and development of own reactor projects.

Also, the PRC government developed the "Energy Security Concept" for five years, which was included in the 12th Five-Year Country Development Plan (2011–2015) announced at the NPC Congress in March 2011.

In accordance with this Concept, all aspects of the development of China's nuclear power industry are under state control. In this regard, decisions on the principles and directions of the development of the industry are made at the highest state level. The center of decision-making on nuclear policy is the State Council of the People's Republic of China. Decisions are developed in the State Commission for Development and Reform based on data coming from the State Energy Bureau.

For the implementation of the decisions taken, mainly two state companies are responsible: the Chinese State Nuclear Energy Corporation (CNNC) and the Chinese Nuclear Energy Corporation of Guangdong (CGNPC). Both companies are building and operating nuclear power plants in China. However, it turned out that the capacity for conversion, uranium enrichment, and production of nuclear fuel is in the hands of CNNC, as well as most of the research centers conducting nuclear power.

As part of the energy security concept until 2015, CNNC had to invest more than 500 billion Yuan ($78 billion) in nuclear power plant construction, which have increased the installed capacity of the nuclear reactor fleet to 40 GW. As a result of the measures being taken, today, China in terms of the number of reactors under construction is considerably ahead of the rest of the nuclear powers, including Russia and India. Moreover, China is actively developing new technologies in the nuclear industry. So, for example, in August 2011, Chinese media reported the creation of its first nuclear reactor of the 4th generation. China Institute of Atomic Energy, the experimental fast neutron reactor, which was developed for the last 20 years, will reduce the amount of radioactive waste. Thus, China became the ninth country to develop a fast neutron reactor that uses uranium 60 times more efficient than a conventional reactor, which will reduce the country's dependence on imports of this feedstock.

Nevertheless, despite some achievements, the nuclear power of the PRC still occupies one of the last places in the country's overall energy balance. According to 2010 data, more than 70% of energy in China is generated by thermal power plants operating on coal and hydrocarbon fuel, about 17.5%—hydroelectric power stations and only 2.3%—nuclear ones. Thus, China, to date, is not included in the list of even ten countries in terms of the number of installed capacity of nuclear power plants. According to the latest data, there are 443 nuclear reactors in 31 countries with a total installed capacity of 377.7 GW. The most extensive stock of nuclear power plants in the world is in the United States (104 nuclear power units) and France.

Also, among Chinese experts on nuclear energy, there is no unity of opinion on such an important issue as a priority type of rector. In particular, the main dispute is the choice between the reactors of the 2nd and 3rd generations. Chinese and foreign researchers have repeatedly noted that the second-generation CPR-1000 reactors under construction and planned by the time of the end of their service life (by 2070–2080) will become obsolete in terms of safety, which would seriously increase the overall risks of the Chinese nuclear facility.

To date, all 14 power units of nuclear power stations in China are equipped with 2nd generation reactors. From this, it follows that at the moment the Chinese government chose the CPR-1000, a second-generation reactor with pressurized water, based on French technological achievements, as the basic type of reactor for its nuclear power industry.

Another significant obstacle to the development of nuclear power is China's lack of advanced technologies in this area. In all of today's nuclear power plants in China, a significant share of technology imported from abroad and mastered with the participation of foreign firms (from Russia, Canada, France, and the USA).

Thus, China does not occupy a leading place in the world nuclear community. However, based on the plans of the State Council of China on large-scale construction of new nuclear power plants, it can be assumed that in the medium term, China will significantly strengthen its position among countries using nuclear energy. At the same time, most likely the accident in Japan has become a powerful catalyst for the process of reviewing the nuclear facilities under construction, on the one hand, and reform of the legislative base, on the other. It is expected that after the appropriate checks the leadership of the State Council will adopt new legal acts regarding the nuclear industry, providing for the expansion of the staff of supervisory authorities, as well as the creation of the supreme supervisory authority for the country's nuclear energy.

15

Future of Nuclear Power Development

Future of Military Atom

HEU-LEU Program: The Triumph of Russian Technology

In the 1990s and later, Russia often launched missiles, as part of a routine training program. After the collapse of the USSR and devastating Perestroika period in the '90s, Russian deterrence systems were in disarray. Russian military secrets were exported in tons. However, even in such terrible conditions, Russians never allowed themselves to get rid of the weapons of the "last resort" (intercontinental nuclear missiles).

Since then, the state of the Russian nuclear deterrent has improved dramatically along with the Russian second-strike capability. Russia built remarkable intercontinental systems (for example, the Topol, the Bulava) designed to repel any predictable threats. Moreover, the fleet of Russian submarines has been significantly modified, which can not only protect but also can attack American submarines and even the American continent by cruise missiles. The Russians now have the latest and most advanced ballistic missiles for submarines.

D. Basu and V. W. Miroshnik, *The Political Economy of Nuclear Energy*,
https://doi.org/10.1007/978-3-030-27029-2_15

Russia and the United States signed in February 1993 an intergovernmental agreement on the disposition of highly enriched uranium (HEU) extracted from nuclear weapons. It should be understood that the nuclear charge gradually deteriorates during storage and its fissile materials due to self-decay are progressively poisoned by the isotopes formed. It became clear that with such an excess of old warheads, they should be disposed of diluted with low-enriched uranium (LEU) and used as fuel in nuclear power plants.

Thus, Russia undertook to down-blend 500 tons of HEU (enough to build 20,000 nuclear warheads) over 20 years, known as the HEU-LEU program (the informal name of the program is Megatons to Megawatts). According to the deal, the two sides agreed that the LEU would be used as fuel by US nuclear power plants.

Thomas Neff, the scientist from MIT Center of International Studies, proposed the idea of down-blending excess stockpiles of weapons as HEU and using the resulting LEU as fuel for nuclear power plants (Bukharin 2000). And this idea was accepted eagerly by signing in July 1991 of the Soviet-US START I Treaty. This agreement mandated a reduction of the two countries' nuclear weapons stockpiles by approximately 5000 warheads apiece (Bulletin of the Atomic Scientists, 1995). Moscow expressed its interest in Neff's proposal due to the difficult economic situation in the USSR at that time. This deal was the prospect of billions of US dollars in hard currency earnings being generated as a by-product of implementing the program. This money was used to support the Russian nuclear industry, which was suffering from a sharp reduction in government funding. Russian talented experts developed the unique methodology which relies on gas-phase dilution by mixing HEU hexafluoride with hexafluoride of slightly enriched uranium. The blend-stock used in that technology is depleted uranium produced by uranium enrichment plants and later enriched to 1.5%. Such a solution had three advantages. First, it reduced the U-234 content to a level deemed acceptable under US standards. Second, it reduced the wastage of separation work. And third, it made it possible to achieve a compromise with the United States regarding the amount of LEU to be bought by USEC, with a right balance struck between the size of the US market and the amount of spare separation capacity in the Russian uranium enrichment industry. The Peaceful Nuclear Energy

Cooperation Agreement between Russia and the United States was signed on May 6, 2008; it entered into force on January 11, 2011 (http://www.state.gov/documents/organization/187744.pdf).

For decades the HEU-LEU Agreement kept the Russian enrichment plants busy producing the slightly enriched blend-stock which was roughly equivalent to the amount of work that would have been necessary if Russia were to supply the United States with the same quantity of LEU produced from natural uranium. According to a preliminary assessment of sales of the natural component to the group of Western companies and the return of some of the natural component to Russia (taking into account the LEU shipments to the United States due before the end of 2013), the overall revenue the HEU-LEU Agreement has generated for Russia could be as high as 17bn USD, with about 13bn USD in hard currency going directly to the treasury. The implementation of the Russian-US Agreement on the disposition of HEU extracted from nuclear weapons has become an effective instrument of irreversible nuclear disarmament. As a party to the Agreement, Russia has converted 500 tons of its HEU stockpiles into LEU that is suitable for use as power reactor fuel.

The importance of the HEU-LEU deal for the United States can be described as follows: for almost 20 years LEU supplies under the Agreement with Russia have accounted for about 50% of the nuclear fuel consumed by the American NPPs. Up to 10% of American electricity is generated from uranium supplied under the HEU-LEU program.20.

The revenue generated by the program, especially in the 1990s, made a substantial contribution not only to the Russian nuclear industry's bottom line but to the Russian treasury as well. The HEU-LEU Agreement has been a useful platform to demonstrate the possibility of using commercial approaches in the implementation of disarmament initiatives. It has also enabled the Russian and American nuclear industries to gain valuable experience of working together, which will facilitate further cooperation in commercial uranium enrichment services.

However, there is no prospect for sales of Russian uranium products in the United States after the completion of the HEU-LEU Agreement in 2013. Russia has no intention to extend the HEU-LEU Agreement; senior ROSATOM executives have made it clear on more than one occasion.

Russia's own nuclear energy industry will require the remaining Russian HEU stockpiles.

Russia Achieved Nuclear Superiority over the United States

The strategic miscalculations of Washington led to a radical change in the military balance of power. Today in Europe, Russia is much stronger than NATO. Henceforth, Russian military domination on the continent is a specific factor.

The US State Department published in September 2014 a report stating that Russia, first time since the collapse of the USSR, had reached parity with Washington in the field of strategic nuclear weapons. Thus, Washington recognized that Moscow regained the status that the Soviet Union achieved by the mid-1970s at the cost of incredible efforts and which was lost after the collapse of the Union due to the betrayal of corrupted Soviet elites.

As follows from the report of the State Department, in 2017, Russia has 528 strategic nuclear weapons carriers, which deployed 1643 warheads, and the United States—794 carriers and 1652 nuclear warheads.

However, it turns out that Russia has strategic nuclear forces (SNFs) that are even more high-tech than the US forces since they provide the final parity for warheads by a substantially smaller number of carriers of strategic nuclear weapons. In the light of the statements of top Russian leaders, by 2020 the SNFs of Russia will be whole, 100%, rearmed with missiles of a new generation; hence the gap between Russia and the United States will only increase.

This breakthrough was made possible by the Treaty on the Limitation of Nuclear Weapons, also known as START III, signed by Dmitry Medvedev and Barack Obama on April 8, 2010 in Prague (entered into force on February 5, 2011), which provides for the reduction of nuclear warheads by 2021 up to 1550 units, and carriers (intercontinental ballistic missiles (ICBMs), submarine-launched ballistic missiles and heavy bombers)—up to 700 units.

That was the first treaty in the strategic area in which Russia managed to achieve significant advantages for itself. In it, for the first time, the Americans pledged to reduce their strategic potential, while Russia was able to enhance it. In addition, it removed from Russia the most essential restrictions in previous START I and START II agreements: the size of the deployment areas for mobile ICBMs, the number of multi-charge ICBMs, and the possibility of creating railway ICBMs.

Portraying Russia as a sworn enemy and believing in the myth of its unattainable military and technological superiority, the United States has driven itself into a trap, the way out of which—in any case in the near and medium term—is not even visible.

And it's not just SNFs. There were many discussions in the United States about so-called "*Wars of the sixth generation*" or in other words the high-precision long-range weapons, which are designed to ensure victory over the enemy without coming into direct contact with his armed forces. Apart from the fact that the concept itself is very doubtful as neither in Iraq nor in Afghanistan, the United States never achieved the victory in this way. However, Russia is here on parity lines, proof of this is in the long-range missiles of a new generation, which would soon be placed on submarines of the Black Sea Fleet and missile ships of the Caspian Fleet.

Soviet Army had approximately 20,000 tanks by the collapse of the USSR. The Americans, in turn, deployed a group of 6000 heavy tanks Abrams in the Allied territory. However, despite this, the aggregate potential of NATO in Europe was still substantially inferior to that of the Soviet Union. And to compensate for this imbalance of NATO strategists were forced to develop the tactical nuclear weapons (TNW).

A study was conducted in NATO in the 1950s about the forces necessary for the bloc superior forces of the USSR and the Warsaw Pact countries. Then the calculations showed that to solve this problem, it is required to have at least 96 divisions. Meanwhile, the costs per just one division exceeded $1 billion (and this is in the prices of that time!). In addition, about 2–3 times more funds were required for the maintenance of such a huge grouping of troops and the creation of an appropriate infrastructure. This burden was clearly beyond the strength of the entire Western economy. A way out was to deploy a group of American TNW (Tactical Nuclear Weapons) on the continent, which was soon done. By the early 1970s, the US TNW arsenal arrived in Europe to protect it from the potential danger of Soviet tanks.

World Strategic Nuclear Weapons: Fatal Miscalculation of West

The West believed that Russia would never rise again, that the possibility of a great war in Europe was reduced to zero and the Russian TNW, along with Russian tanks, would in due course disintegrate. The reality is different.

NATO countries currently have only 260 TNW. Two hundred bombs with a total capacity of 18 megatons are in the United States. They are deployed in six airbases in Germany, Italy, Belgium, the Netherlands, and Turkey. Another 60 nuclear bombs are in France. Russia today, according to the most conservative estimates, has five thousand units of TNWs of different classes—from the "Iskander" to military aircraft and artillery. The United States cannot change this imbalance: The rest of the "Cold War legacy"—both tactical nuclear shells and land-based missiles, and nuclear warheads of the Tomahawk sea-based cruise missiles—they destroyed.

In order to understand how it so happened that Russia, which has lost the Cold war, is superior to NATO, it is necessary to turn to the history of the issue. It was believed that by early 1991, the USSR had approximately 20–22,000 units of TNW.

It is important to note that this impressive military arsenal of Russia was the result of forty years of intense arms race, which, incidentally, was not initiated by the "totalitarian" USSR, but quite democratic and liberal by the United States, which already in the early 1950s started developing and testing various types of TNW. By the way, the Soviet leadership repeatedly invited the Western leaders to begin negotiations on the reduction of TNW. NATO, for a long time, has stubbornly rejected all proposals of the USSR on this issue. The situation radically changed during Perestroika in USSR. Then Washington considered that it was necessary to use the moment to weaken and disarm its main geopolitical enemy maximally.

In September 1991, US President George W. Bush proposed an initiative to reduce or even eliminate certain types of TNW in the United

States. Gorbachev, in turn, also announced plans for a radical reduction of similar weapons in the USSR. Subsequently, these plans were developed in the statement of the President of the Russian Federation Boris Yeltsin named "*On Russian policy in the field of arms limitation and the reduction*," January 29, 1992. It stated that Russia had stopped the production of nuclear artillery shells and warheads for land-based missiles, and all stocks of such warheads would be destroyed. Russia will remove all TNW from surface ships and multi-purpose submarines and eliminate one-third. The facts now are obvious: Russia kept the promise, the United States did not.

Meanwhile, in Russia, experts quickly agreed that in the geostrategic situation that developed after the collapse of the USSR, the massive reduction and destruction of TNW is unacceptable. This led to the fact that now—after Russia overcame the inertia of disintegration and proceeded to the systematic restoration of its power, and the West, lulled by sweet dreams about the liberal idea of "end of history," castrated its armed forces to the point where they are capable of leading only colonial wars with a weak, technically backward adversary—the balance of forces in Europe has radically changed in Russia's favor.

In December 2010, Assistant Secretary of State for Verification, Compliance, and Implementation of Agreements, Rose Goethemuller, sounded the alarm: "*Russian tactical nuclear systems have more than ours, and Congress strongly recommends that these issues be dealt with … The next step should be the reduction of tactical nuclear weapons*" (http://www.state.gov/r/pa/ei/biog/121630.htm).

Europeans showed a great deal of activity in the same year in the face of the heads of foreign affairs agencies of Poland and Sweden, insolently demanding from Russia the unilateral creation of two nuclear-free zones (a) the Kaliningrad region Kola Peninsula (the priority areas deployment of Russian TNW, including the main base area of the Baltic) and (b) Northern fleets (another priority area deployment of Russian TNW).

Since then, Americans have repeatedly offered Russia a flawed way to solve the problem of TN, stubbornly insisting on the development of an agreement on eliminating the inequality in the stocks of TNW. They even tried to determine the entry into force of the SALT-3 treaty by the beginning of negotiations on TNW. Thus, in accordance with the Senate Amendment Sen. Lemie (Amendment 4/S.AMDN.4908), the final entry into force of SALT-3 should take place only after the Russian side agreed to negotiate a so-called "*elimination of imbalance*" in TNW of Russia and the United States (https://www.congress.gov/111/cdoc/tdoc5/CDOC-111tdoc5.pdf).

On February 3, 2011, Barack Obama in a letter sent to a number of key senators, said "*the beginning in the near future of negotiations with Russia on eliminating disparity between tactical nuclear weapons of the Russian Federation and the United States and reducing the number of tactical nuclear warheads in a verifiable way, are needed*" (https://www.armscontrol.org/act/2011_01-02/NewSTART). In 2012, Putin returned to the Kremlin and the hopes of the West to "breed" Russia, inciting it to unilateral disarmament, failed.

The price of this failure became apparent during the Ukrainian crisis: The West lost its former military superiority over Russia. There are no ways to rectify this disparity either in Washington, in London, in Berlin, or Paris. Henceforth, Russian military superiority in Europe is an indisputable factor.

Future of Peaceful Atom

Events and Trends in the Nuclear Power Industry

An analysis of recent scientific work has shown that scientists agree that the future of nuclear power is unthinkable without fast-neutron reactors that will reprocess depleted nuclear fuel. They allow more efficient use of uranium reserves to solve environmental problems. And they are even safer than the nuclear power units currently under construction. The study was aiming to consider the prospects for the development of nuclear energy in the context of global trends in the development of nuclear technology.

Since the beginning of the 2000s and to this day, nuclear power engineering has been at the stage of recovery, mainly due to the large-scale construction of nuclear power plants in the countries of the Asia-Pacific region. Accelerated closure of obsolete NPPs is observed in the world, and requirements for state and international regulation of the safety of industrial facilities are being tightened. However, the accident at the Fukushima

Daiichi NPP in Japan made specific adjustments, which negatively affected the industry's projected growth rates. Undoubtedly, fossil fuels (oil, natural gas, and coal) still satisfy most of the world's needs, because no other energy source can compare with them in accessibility, efficiency, and scale of production. However, among non-fossil fuels, it is precisely the proportion of atomic energy in energy consumption that, although slowly, is growing.

According to the forecasts of the World Nuclear Association in 2016, the global nuclear power capacity will increase by 26.7% (to 494 GW) in 2030 and by more than 40% (546 GW) in 2035 (with a high scenario). With a low scenario, nuclear power capacities in 2030 will remain at about the current level. It is assumed that in 2050 the nuclear potential will be about 964 GW, providing 24% of the global electricity supply. The growth of nuclear power is possible due to the accelerated transition to standard reactors of the III and IV generations, as well as to fast-neutron reactors. That will solve the problems of both supplying countries with uranium ore and spent nuclear fuel, improving the economic performance and safety of nuclear power plants. Generation III reactors are characterized by lower capital costs and shorter construction times; large capacity utilization and long service life (typical period is 60 years); simple and reliable design, easy to maintain and less vulnerable to operational problems; reduced probability of accidents with the melting of the core; minimal environmental impact; more significant burnout of the fuel to reduce waste and fuel requirements; using scavengers that burn to extend the life of the fuel cells. Reactors of the fourth generation will be economically efficient, safer, will produce less long-lived radioactive waste and will meet the requirements for the non-proliferation of nuclear technologies and materials.

Research and development of IV generation reactors are carried out in the framework of the Generation IV International Forum program, in which Argentina, Brazil, Great Britain, Canada, South Korea, South Africa, the USA, France, Switzerland, Japan, and Euratom participate, as well as a part of the International Project the development of innovative nuclear reactors and fuel cycles.

Today, the number of requests to the International Atomic Energy Agency (IAEA), especially from developing countries, is growing rapidly to provide them with "technical conditions" for the creation of energy systems, where nuclear power plays an important role. Today, only eight countries have nuclear potential. Fifty-five countries have 245 civilian experimental reactors; more than 1/3 of them are in developing countries. As of September 1, 2016, 31 countries have 447 industrial nuclear reactors with a total installed capacity of over 390,000 MW. That is more than three times the total energy capacity of all the resources of France and Germany. About 60 additional nuclear reactors are under construction, which is equivalent to 16% of existing capacity, while more than 160 are at the proposal stage for construction, which is equal to almost half of the current capacity. In 16 countries, a quarter of electricity is generated by nuclear power plants, in France—three quarters, in Belgium, the Czech Republic, Finland, Hungary, Slovakia, Sweden, Switzerland, Slovenia, and Ukraine—about one-third and more. Nuclear power plants generate over 30% of electricity in South Korea and Bulgaria. In the United States, Great Britain, Spain, Romania, and Russia, almost one-fifth of the energy is nuclear. In Japan, more than one-fourth of the electricity was generated by nuclear power plants, and after the accident at the nuclear power plant, the country returns to this level. In Italy and Denmark, only 10% of electricity is of nuclear origin. Today, of all operating NPP reactors in the world, almost half are in operation.

The combination of a number of factors, internal and external, is a source of severe problems for the nuclear power industry in some of the most developed markets. Despite this, the need for reliable and predictable supply of "clean" electricity, which is provided by nuclear power, has never been so high, which was reflected in the growing number of new NPP construction programs implemented in several parts of the world. At the end of 2017, out of 448 operating units, more than half were located in the United States and Europe, where, despite the importance of nuclear energy, the closure rates continue to outpace the replacement rates of outgoing capacity.

The strength of the Russian nuclear industry is manifested in its dominance in the export markets for the construction of new NPPs. Currently, the country's nuclear industry participates in projects for the construction of new power units in the Republic of Bangladesh, China, the Republic of Belarus, Hungary, India, Iran, and Turkey. In varying degrees, Russia acts as a potential investor in projects in Algeria, Bolivia, Brazil, DRC, Egypt, Indonesia, Jordan, Kazakhstan, Nigeria, the Philippines, Saudi Arabia, South Africa, Sri Lanka, and Tajikistan.

The difficulties faced by the American nuclear industry had a positive effect—they made it possible for those who make decisions in the states and at the federal level to realize the value of nuclear energy. In April 2018, the state of New Jersey introduced the Zero Emission Certification Program (ZEC), under which nuclear power plants will receive compensation for carbon-free power generation and a contribution to fuel diversification.

The peak of the development of nuclear energy has passed. The maximum share of electricity generated by nuclear power plants in global electricity production (17.5%) was reached in 1996. Since then, this share has decreased, and today is only 10.5%. The absolute peak of annual output (2660 TWh) was reached in 2006, in 2016 it was made less—2476 TWh.

The Big Five—the United States, France, China, Russia, and South Korea—produce 70% of the world's atomic electricity. The share of the United States and France is 48% in total. Ten reactors were commissioned in 2016, half of which in China. The average construction period was 10.6 years (excluding the American Watts Bar-2, the longest construction project, 7.1 years).

The oldest reactors are located in Switzerland; their average age is 42.2 years. In the world, this figure is 29.3 years, in the EU—32.4 years. Recall that the Swiss at a recent referendum decided not to build more nuclear power plants. Therefore, soon, the world nuclear power industry will lose one of its customers. Annual global investments in nuclear power are many times less than in solar and wind generation (see the graph where the red line is investments in nuclear power).

Electricity generation based on wind energy in 2016 was 948 TWh more than 20 years ago (1997), solar energy—332 TWh. Last year, nuclear power produced a total of 212 TWh more than in 1997. In China, since 2012, wind power generates more electricity every year than nuclear power. From 2000 to 2016, 451 GW of new capacities were built in the global wind power industry, 301 GW in the solar power industry, and 36 GW in the nuclear power industry.

Electricity generation based on wind energy in 2016 was 948 TWh more than 20 years ago (1997), solar energy—332 TWh. Last year, nuclear power produced a total of 212 TWh more than in 1997.

In most nuclear powers (Brazil, Germany, Japan, Great Britain, Spain, India, Mexico, and the Netherlands), more electricity is generated based on renewable energy sources (excluding hydropower plants!) than nuclear power plants. In Europe, there is a decrease in the installed capacity of nuclear power plants and an absolute decrease in their output.

The decline in the cost of wind and solar technology in recent years has led to a sharp drop in prices for solar and wind electricity. In the most favorable, from a climatic point of view, regions, solar energy produces the cheapest electricity in the world. The cost of a megawatt-hour at a depreciated nuclear power plant in the United States is $35.5. If we talk about new construction, the calculated cost price (LCOE) megawatt-hour in American nuclear energy is already around $150.

In the coming years, new renewable energy sources (wind and solar energy) will be ahead of the peaceful atom not only in terms of installed capacity but also in output (in 2016, the total share of wind and sun in world generation was approximately 6.5%). Nuclear power plants are too expensive, their construction times are very high, the problems of radioactive waste disposal are not solved, and political decision-making processes about choosing a peaceful atom as a source of electricity are extremely complex. Nuclear energy, alas, does not fit well into today's rapidly changing world.

In 2017, in Russia, the share of nuclear power plants (NPP) accounted for 17.8% of all electricity produced in the country. In the world today, according to estimates of the IAEA, there are a total of 188 nuclear power plants, which employ 450 reactors. More than a third of them are concentrated in only two countries—the United States (99 reactors) and France (58). For comparison, in Russia—37 operating reactors at ten nuclear power plants. Accordingly, the share of the United States and France accounts for almost half of all electricity produced in the world at nuclear power plants. In general, the top ten countries account for 85% of its production.

Any country itself determines its own energy balance, that is, what types of fuel to use for the production of electricity, which should be based on their capabilities. If we bear in mind the overall energy balance of the world, then the share of nuclear energy in it is now less than 11%. In general, 30 countries of the world already possess such energy. And the share of nuclear power plants in total electricity generation in these countries may differ tenfold.

Most operating reactors are concentrated today in the European Union—130 in 14 countries of the region. The share of nuclear power plants in the total energy balance of the European Union decreased from 30% in 1990 to 26% in 2016. However, with regard to nuclear energy, the policies of individual European countries can vary greatly. That was particularly noticeable after the accident at the Japanese "Fukushima" in 2011. We will give just a few examples.

Although in Europe the share of nuclear power plants today is 53% of all electricity produced without the use of hydrocarbons, nevertheless, in the next 12 years, the total capacity of decommissioned reactors (for technological or political reasons) will exceed the total capacity of new reactors under construction, state Nuclear Association (WNA). So far, only four new reactors are being built in Europe—in France, Finland, and Slovakia (2). Construction of another 24 is planned, of which 11 are in the UK, 6 in Poland, two each in the Czech Republic, Hungary and Romania. However, in Russia 5 new reactors are already being built here, and 26 more are planned.

In general, throughout the world, according to the IAEA, the number of operating reactors has changed little over the past 20 years. If in 1997 there were 441, then in 2016—450. At the same time, 56 new reactors are being built in the world at once, that is, six times more than they have been added over two decades, of which 18 are in China alone. And according to the current forecast of the International Energy Agency, by 2030, China is already seriously ahead of the United States in view of the amount of electricity produced in the country at nuclear power plants.

The total amount of electricity produced in the world at power plants of all types will double by 2050, according to the current IAEA forecast. However, the share of nuclear energy in the total electricity generation of the world, according to the same forecast options, will either halve or increase by only one-third.

Reference

Bukharin, O. (2000, June). *Helping Russia Downsize Its Nuclear Complex: A Focus on the Closed Nuclear Cities*. Princeton: Princeton University Press.

Conclusion

Future of Nuclear Power: Scientific and Technological Development

In the twenty-first century, science, knowledge, information, and human resources become the primary productive forces. All this forms a new technological structure. A full scientific complex was created in the USSR, and the concept of the great triad was formulated: fundamental science, applied science, and personnel training. Now people worldwide need to focus on the components of this triad to launch an overtaking rather than a catch-up development, create a fundamentally new nuclear technology to lead their country to success in the future.

Unfortunately, over the past quarter-century, much has been lost in nuclear technology development worldwide, and only the Russian nuclear industry managed to preserve the highest scientific and technical potential, with the global leader in the nuclear industry as Russian state corporation ROSATOM.

In order to gain technological sovereignty, it is necessary to continue training personnel, improve their level of education, and set the right

D. Basu and V. W. Miroshnik, *The Political Economy of Nuclear Energy*, https://doi.org/10.1007/978-3-030-27029-2

tasks for them. To launch the reindustrialization of the economy, to plan scientific and industrial development need for decades to come. Sovereign development and creation of our own technologies should become our national idea, since only they will provide our citizens with a high quality of life, and the country—security and intellectual leadership in the future. The acquisition of fundamental scientific knowledge and the creation of modern technologies require the concentration of enormous resources: scientific, informational, financial, technological.

The objectives of the nuclear industry are not limited to the improvement of reactors and the commissioning of new NPPs. For example, ROSATOM enterprises have dual-use technologies. Nuclear-medicine and gas lasers, devices for the detection of explosives and radiation protection, the creation of composite materials and the processing of nuclear waste are also high technologies that must be improved in order to remain among the leading countries.

The Future of Nuclear Power

The decisive factors of the investment climate in the nuclear power industry are the state's attitude and the degree of its intervention in the industry. These factors affect all others, but the vectors of their impact are directly opposite. It is desirable to provide some examples.

First of all, with a high level of protectionism, guarantees of return on investment are high, which attracts private investors, ensuring a secure return on investment. However, if in moderate cases such protectionism increases the investment attractiveness of the industry (for example, in the UK), then in the case of "nuclear absolutism," the path for private, especially foreign investments in nuclear energy may be closed or the competition severely restricted. This leads to a narrowing of the choice of technology, rising prices and limiting the number of investors, and in some cases—the volume of investment. Examples can be found in Russia, South Korea, and other countries.

So, the degree of state participation in the nuclear industry does not in itself determine the quality of the investment climate; it is a double-edged sword. The intervention of the authorities may be for good, and maybe to the detriment. There are examples of the successful development of nuclear power both under the protectionist (Russia, France, and South Korea) and under the market model of the industry (the USA, Finland, and Germany in the past).

The very model of the industry is not static; it often evolves, and with it—and the investment climate. Regardless of how the evolution of the industry began, it can lead to entirely different results. Conservative protectionism sometimes leads to the dynamic development of nuclear power and its successful integration into a competitive market (Russia). At the same time, liberalism with the most-favored-nation regime for investment suddenly turns into an authoritarian dictatorship of the state, the "ice age" begins in the investment climate, and the planned liquidation of the industry (Germany, Belgium, Switzerland, and Sweden in part).

Such a finale is essentially "nuclear absolutism inside out": external signs of the market investment climate persist (absence of state monopoly, competition in all segments of the industry, liberal market, etc.), but in fact, the government directly sets the plan for the development of nuclear energy. That is, the discriminative investment climate is the last stage of the evolution of the industry.

However, the negative position of the state is not always the main obstacle to nuclear energy. In some countries and regions, the combination of natural and market conditions makes the use of nuclear energy economically inexpedient. For example, in Norway, New Zealand, Austria, the abundance of cheap and well-developed hydropower resources has always questioned the feasibility of the appearance of nuclear power plants. The same factor in several countries where nuclear power plants exist (Brazil, Sweden, Canada, Switzerland), creates attractive alternatives to the existing nuclear power industry, hampering the development of the industry.

Sometimes the state, even without objective reasons, creates prohibitive barriers to nuclear power, despite the undoubted economic feasibility of the nuclear power plant for a given country. This is the case, for example, in Italy—the most volatile country in Europe, which does not have enough of its own energy sources to meet the needs of the fourth largest EU economy. Italy tried to develop a whole range of nuclear technologies, but after the accident at the Chernobyl nuclear power plant, Italy turned off the projects. The attempt to revive nuclear power at the beginning of this century "stumbled" about the accident in Fukushima: the national referendum that followed after it forced the government to confirm the denuclearized status of Italy.

Absolute insensitivity to public opinion and advances in recent technology can lead to an overly simplistic, voluntary order of approving nuclear projects, that is, to some formalization of nuclear oversight. This ultimately weakens the nuclear safety system, dulls the vigilance of supervision and can lead to dramatic consequences. Something similar, apparently, took place in Chernobyl in Ukraine, Fukushima in Japan or The Three Miles Island in the United States.

However, the coin has a downside: a high degree of liberalization not only of the market but of society can do a disservice to nuclear energy, forcing the country to choose a far from optimal energy strategy. In states with a developed democratic tradition, the decision to use atomic energy and the fate of multi-billion investments is sometimes determined by the mood of the voters, most of whom are vaguely aware of the real advantages and dangers of nuclear power. These moods depend on random events and targeted manipulations of interested and social groups. One such incident, which occurred five years ago in Fukushima, was predetermined a sharp reversal in the energy sector of a number of countries.

Thus, the investment climate for nuclear energy is very different in different countries, regardless of their other characteristics, and is changing at times in the most unexpected way. The role of the state in this is great, as in no other sphere. But there is no direct correlation between the conditions of nuclear power plant development and the state "dictatorship" or "democracy" in the nuclear industry. Nuclear energy is a rare area where both can produce equally good results.

The economic situation, equal conditions for business and liberalization of the markets do not outweigh this factor: sometimes in countries with a developed and prosperous economy nuclear power is in crisis, while in some developing countries it is progressing

The analysis of the history of nuclear science and technology makes us think about the prophecy of the great Russian scientist Vladimir Vernadsky a hundred years ago. The nuclear community must find a way out of this contradiction: nuclear technology has not yet brought humanity into the golden age of solving energy problems (with its current contribution to the full energy balance of about 5%), but it has allowed humans to create nuclear weapons with potential capable of destroying humanity.

Also, the development of aerospace technologies using nuclear and electronuclear rocket engines is important for the future. On rocket engines that use the chemical energy of combustion of fuel components, it will be difficult and extremely inefficient to master deep space or plan flights to Mars. To accomplish the plans of the Moon discovery, it is necessary to have at least 6 MW of power in orbit; to successfully accomplish the Mars mission, it needs about 25 MW of power. This level of energy can only provide nuclear technology. Further strategic development of the nuclear industry is not just a means of achieving technological sovereignty. That is the generation of new knowledge about the structure of matter and its properties. Thus, the creation of the scientific and engineering potential of the world is still ahead.

Appendix

Natural Uranium

Considering the reserves (proven recoverable reserves) of nuclear fuel for one comparatively rare isotope of uranium (U235, which in natural uranium is only about 0.7%), we have the equivalent to about 90 billion tons of oil or only half as much as the oil reserves on our planet. However, in addition to U235, we also have thorium and the isotope U238. Together with them, the aggregate energy potential of the world's available nuclear fuel reserves exceeds the energy potential of world oil reserves by about 150 times.

The technology for the production of nuclear energy has so far been mastered only for an enriched mixture of the uranium isotopes (the content of U235 in this mixture needs to be increased from natural 0.7% to 3–4%). Regarding natural uranium, in 2011 the world reserves are, according to the International Atomic Energy Agency (IAEA), about 6.5 million tons. Most important countries possessing these reserves (in thousands of tons) are as follows:

D. Basu and V. W. Miroshnik, *The Political Economy of Nuclear Energy*, https://doi.org/10.1007/978-3-030-27029-2

Most of these leaders of the world nuclear power industry (France, South Korea, Ukraine, Great Britain, Sweden, and Spain) have almost no reserves of their own, while the reserves of the United States and China do not have enough nuclear reserves to create fuel for self-sufficiency. Therefore, for direct or indirect control of existing large uranium deposits, there is an acute possibility of war. There were tribal conflicts, civil wars and coups d'état in Gabon and Niger. Niger has been for many years the main supplier of uranium for the nuclear power industry in France, and recent years also for China. Another war is in Mali. In the north of this country, large uranium deposits were discovered, to which China, France, and the USA immediately showed great interest. Tuareg separatists aimed at creating the independent state of Azawad, as well as the radical Islamists of Al-Qaeda and the Ansar-Din movement, were trying to establish an Islamic state based on Sharia in Mali. US interest in Mali is largely due to uranium. The former head of state Amadou Toure is unacceptable for the United States precisely because he accepted the Chinese into the country. Current Prime Minister of Mali Sheik Diarra (an astrophysicist who worked in the US NASA and then at the Microsoft office in Africa) is the "US candidate" for the presidency of Mali.

However, this is not restricted to Africa. Kyrgyzstan is another area of conflict. In 2010 there were ethnic clan conflicts over the issue of future control over Kyrgyz uranium reserves. Sometimes this "war for uranium" is quiet. In Kazakhstan companies from Japan, South Korea, and China are operating together. Some foreign companies purchased shares in a number of Kazakhstani uranium deposits, and others (primarily, Chinese ones) concluded long-term contracts for the supply of Kazakhstani uranium; besides there are Canadian company Uranium One, US General Atomics and Russian ROSATOM.

According to the latest (2012) minimum forecast of the IAEA, by 2030 the installed capacity of nuclear power plants in the world will increase from the current 370–447 GW. The planned capacity of the NPP in 2030 will reach 691 GW. Accordingly, the consumption of enriched uranium will also grow. Thus, the one who extracts and enriches uranium better and cheaper simultaneously gets the opportunity to dominate the NPP construction market.

References

Basu, Dipak. (1981). *Future energy Policy of the UK.* London: Macmillan.

Brodov, V. (1984). *Indian Philosophy in Modern Times.* Moscow: Progress Publishers.

Bukharin, O. (2000, June). *Helping Russia Downsize Its Nuclear Complex: A Focus on the Closed Nuclear Cities.* Princeton: Princeton University Press.

Coriou, H. (1959). *Colloque de Metallurgie sur la Corrosion* (p. 161). Amsterdam: North Holland Publishing.

Dasgupta, S. (1922). *History of Indian Philosophy.* Cambridge: Cambridge University Press.

Energetics of the XXI Century. Newsletter No. 1, April 2008. Institute for Innovative Energy Russian Research Center "Kurchat*ov Institute".*

Freeman, K. (1948). *Ancilla to the Pre-Socratic Philosophers.* Cambridge, MA: Harvard University Press.

Gates, R. (2010, May 13). The Case for the New START Treaty. *Wall Street Journal.*

Holloway, D. (1994). *Stalin and the Bomb. The Soviet Union and Atomic Energy 1939–1956.* New Haven: Yale University Press.

D. Basu and V. W. Miroshnik, *The Political Economy of Nuclear Energy,*
https://doi.org/10.1007/978-3-030-27029-2

Kramnik, I. (2010, December 12). Nevsky and Novomoskovsk: Two Submarines for Putin. *Sputnik News*.

Kristensen, H. K., & Norris, R. (2010). Chinese Nuclear Forces 2010. *Bulletin of Atomic Scientists, 66*(6), 134–141.

Kristensen, H. K., & Norris, R. (2018). Russian Nuclear Forces, 2018. *Bulletin of the Atomic Scientists, 74*(3), 185–195.

Lennox, D. (2000, January 12). China's New Cruise Missile Programme 'Racing Ahead'. *Jane's Defence Weekly*, p. 12.

Markova, A. N. (2002). The History of the World Economy. Moscow: UNITI.

McGwire, M. (1991). *Perestroika and Soviet National Security*. Washington: Brookings Institution Press.

Nuclear Posture Review Report. (2010, April). Washington, DC.

OECD. (2008). *Uranium 2007: Resources, Production and Demand*. OECD Nuclear Energy Agency and International Atomic Energy Agency. OECD: Paris. ISBN 9789264047662.

Overy, R. J. (1992). Air Power and the Origins of Deterrence Theory Before 1939. *Journal of Strategic Studies, 15*(1), 73–101.

Radhakrishnan, S. (1923). *Indian Philosophy*. Oxford: Oxford Univeristy Press.

Russian Armed Forces Receive Another Set of Iskander Tactical Missile System. *TASS*, June 28, 2016.

Saunders, P., & Yuan, J. (2000). China's Strategic Force Modernization: Issues and Implications for the United States. In M. Barletta (Ed.), *Proliferation Challenges and Nonproliferation Opportunities for New Administrations* (pp. 40–46). Monterey: Monterey Institute of International Studies.

Schneider, M. (2013, July 31). *Russia's Noncompliance with Arms Control Obligations*. Gatestone Institute.

SIPRI 2010 Yearbook. (2011). Moscow: IMEMO RAN, pp. 373–377.

Vernadsky, V. (1954). *Collection of the Selected Research*. Moscow: Publishing House of the Academy of Sciences of the USSR.

U.S. Department of Defense. (2012). *Annual Report to Congress: Military and Security Developments Involving the People's Republic of China*.

U.S. Department of Energy and U.S. Department of Defense. (2008, September). *National Security and Nuclear Weapons in the 21st Century*, p. 8.

Wallerstein, I. (1974). *The Modern World-System: Capitalist Agriculture and the Origins of the European World-Economy in the Sixteenth Century*. New York: Academic Press.

Wells, H. G. (1914). *The World Set Free*. London: Macmillan.

Wohlstetter, A. (1959). The Delicate Balance of Terror. *Foreign Affairs., 37*(2), 211–234.

References in Russian language

Арбатов А.Г. Оборонная и достаточная безопасность. М., 1990.

Вернадский В. *И.* Избр. соч. М.: Изд-во АН СССР, 1954. Т. 1. С. 623.

Военно-технический прогресс и вооруженные силы СССР: Анализ развития вооружения, организации и опыта действия. Под ред. Генерал-лейтенанта М.М. Кирьяна. М.: Воениздат, 1982.

Вознесенский Н., 1948, Военная экономика СССР в период Отечественной войны: Госполитиздат.

Кокошин А.А. Обеспечение стратегической стабильности в прошлом и настоящем. М.: КРАСАНД, 2009.

Малиновсий Р.Я, Бдительно стоять на страже мира. М.: Воениздат, 1962; Иванов С. Советская военная доктрина и стратегия // Военная мысль. 1969. № 5.

Мерцалов А.Н., Мерцалова Л.Н. Сталинизм и война. М.: Родник, 1994.

Павловский И. Сухопутные войска Советских вооруженных сил // Военная мысль. 1967. № 11; Соколовский В., Чередниченко М. Военная стратегия и ее проблемы // Военная мысль. 1968. № 10.

Трухановский В.Г. Английское ядерное оружие (историко-политический аспект). М.: Международные отношения, 1985.

Пономарев, Н., Степной, Н. «Роль атомной энергетики в структуре мирового энергетического производства XXI века». Электронный журнал «Экологические системы», № 8, август 2006.

Федоров В., Мосалев В. Подводные силы ВМС КНР // Зарубежное военное обозрение. 2010. № 7. С. 52-61.

ЭНЕРГЕТИКА XXI ВЕКА. Информационный бюллетень № 1, апрель 2008 г. Институт инновационной энергетики Российский научный центр «Курчатовский институт».

Марковой, А.Н., 2002, Источник: История мировой экономики / Под ред. Г.Б. Поляка, ЮНИТИ, р. 307-315.

TASS, March 26, 2015

Interfax, December 19, 2012

Internet Reference

http://www.abareconomics.com/interactive.

http://www.armscontrol.org/factsheets/pniglance.

https://www.armscontrol.org/act/2011_01-02/NewSTART.

http://www.atominfo.ru/news/air087.htm.

http://www.atomenergoprom.ru/ru/nuclear.

http://www.atomenergoprom.ru/ru/nuclear/.

http://www.authorstream.com/Presentation/Reva-20059-InzhechikBaksan07.

https://www.congress.gov/111/cdoc/tdoc5/CDOC-111tdoc5.pdf.

http://en.cgnpc.com.cn/.

http://en.cnnc.com.cn/.

https://en.topwar.ru/91488-o-masshtabah-lyudskih-poter-cccr-v-velikoy-otechestvennoy-voyne.html.

http://www.fondsk.ru/news/2015/05/06/ekonomicheskaja-cena-pobedy-sssr-v-velikoj-otechestvennoj-vojne-33176.html.

www.frstrategie.org, 2019.

https://fas.org/wp-content/uploads/2015/10/Nov-Dec-Pakistan; http://vernadsky.lib.ru.

https://www.gatestoneinstitute.org/about/.

Files/69/4_Grachev.pdf.

http://www.iaea.or.at/programmes/a2/.

https://www.iaea.org/PRIS/WorldStatistics/OperationalReactorsByCountry.aspx.

http://www.metalbulletin.ru/analytics_articles.php?id=2859.

http://metaltorg.ru/analytics/publication/?id=2643.

https://www.nkj.ru/archive/articles/2804/.

Olelinikov, P. V. (2019). https://www.nonproliferation.org/wp-content/uploads/npr/72pavel.pdf.

https://nsarchive2.gwu.edu/.

http://nuclearno.ru/text.asp?12857.

http://www.rosenergoatom.ru/rus/press/faq/index.wbp?page=3.

https://www.rulit.me/books/dark-sun-the-making-of-the-hydrogen-bomb-read-253578-7.html.

http://russiancouncil.ru/en/inner/?id_4=7153#top-content.

http://www.sipri.org/yearbook/2011/07.

http://www.state.gov/documents/organization/187744.pdf.

http://www.state.gov/r/pa/ei/biog/121630.htm.

http://www.tvel.ru.

http://www.uxc.com/.

http://www.world-nuclear.org/info/inf104.html?terms=reactors+forecast.

http://vernadsky.lib.ru/.

http://www.vniief.ru.

Index

A

Alpha and beta radiation of uranium nuclei 11
Alpha rays 65, 66
American military doctrine 47
Atom 1, 2, 9, 11, 18, 43, 47, 61–69, 75, 79, 142, 143, 178, 182, 184, 187, 202, 206, 216, 219, 221, 231, 258
Atomic liberalism 98
Atomistic theory 61, 62
Atom smashing 75

B

"Bohr atom" 11
BN-600 reactor 10, 19
Bose–Einstein 43, 216
BWR-type reactor 19, 149, 169, 185, 198, 205

C

Caribbean crisis of 1962 50
Carpet bombing 2, 83
Cavendish Laboratory 75, 141
Chelyabinsk-40 82, 84
Chernobyl disaster 8, 15
Chinese military doctrine 233
Clausewitz's formula 54
Cold war 3, 14, 21, 50, 79, 81, 84, 85, 179, 180, 195, 196, 252

F

Fast-energy power reactor 19
Fat Man 81
Fog of war 50, 51
Fukushima disaster 4, 8, 15, 19, 153, 198–200, 202, 204, 206, 208, 209, 254, 259, 264
Fundamental law of atoms 62, 63, 68

D. Basu and V. W. Miroshnik, *The Political Economy of Nuclear Energy*,
https://doi.org/10.1007/978-3-030-27029-2

G
Gamma rays 66

H
Hafnium 68
Helium-3 11
HEU-LEU Agreement 175, 177, 249
High-enriched uranium (HEU) 20, 73, 79, 175, 177, 240, 248–250
Hiroshima 3, 45, 46, 49, 79, 81, 197
Hydrogen bomb 84, 85, 87–90

K
Kyoto Protocol 38

L
Liquid helium 11
Liquid Metal Fast Breeder Reactor (LMFBR) 172
Little Boy 81
Little Thing 78
Low-enriched uranium (LEU) 175, 177, 248, 249

M
Manhattan Project 45, 71, 76
Mendeleev's periodic system of elements 64
Mendeleev's periodic table 62, 67, 69
Missile Technology Control Regime (MTCR) 194
Mixed oxide fuel 123, 204

N
Nagasaki 3, 45, 46, 49, 79, 81, 197
Natural uranium 8, 9, 20, 43, 72, 73, 225–227, 230, 249, 267
Natural uranium—uranium-238 11, 43, 72
Neutron 9–13, 18–20, 36, 43, 44, 65, 71–75, 90, 135, 137, 143, 144, 147, 152, 171–173, 183–185, 206, 219, 227, 232, 233, 243, 244, 255
Neutron-induced fission 11
Neutron toothpaste 75
Non-Proliferation of Nuclear Weapons Treaty (NPT) 17, 25, 26, 53, 54, 190, 193, 207, 217, 219, 222, 223, 225, 231, 232
Nuclear club 26, 193
Nuclear countries 23–27, 30, 32, 33
Nuclear powers 1, 2, 4, 7, 11–14, 16–21, 26, 27, 47, 48, 57, 85, 195, 225, 244, 258
Nuclear protectionism 98
Nuclear states 25–28, 33, 47
Nuclear umbrella 197
Nuclear World (NM) 23–28, 30–33
Nuremberg Trials 1946 83
NWM intermediate-range nuclear weapons 195

O
Obninsk 2, 9, 84, 90, 91, 173, 185, 186

P
Paris Climate Agreement 38
Plutonium A-1 82
Potsdam Conference 78
Proton 11, 75, 89, 91, 143

Prototype Fast Breeder Reactor (PFBR) 225

Q

Quebec Agreement 76

R

Radioactivity 10, 11, 15, 65, 69, 70, 137, 138
Radium Commission at the Physics and Mathematics Department of the Russian Academy of Sciences 10
Renewable Energy Sources (RES) 38, 106, 111, 112, 117, 121, 125, 154–157, 170, 199, 200, 242, 258
ROSATOM 18, 20, 91, 98, 102, 113, 114, 120, 179, 182, 183, 186, 200, 230, 249, 261, 262, 268
Russian military doctrine 3, 47, 56

S

SALT-1 (1972) 57
SALT-2 (1979) 57
Shale revolution 38
SMRBMs 195
Soviet nuclear bomb 79, 87
Soviet nuclear weapons 11
START I Treaty 248, 251
START II Treaty 251
START III Treaty 159, 161–164, 188
Strategic Missile Force includes three missile armies (RA) 192
Strategic Missile Forces (SMF) 188, 191
Strategic Nuclear Forces (SNF) 47, 52, 55, 57, 164, 165, 167, 188, 191–193, 195, 234, 237, 238, 241, 250, 251
Superfluidity of liquid helium 11
Superstrong magnetic fields 11

T

Tactical Nuclear Weapons (TNW) 47, 56, 166, 189, 251–254
Technological sovereignty 35, 261, 265
Thorium 20, 74, 75, 219, 223, 224, 227, 267
Three Mile Island accident 169
Tritium 11
Tsar-bomb 49
Tube Alloys Project 76
TVEL 18, 118, 120, 175, 176

U

Unthinkable plan 3
Uranium-235 9, 13, 18, 20, 43, 44, 71–74, 136, 143, 144, 146, 152, 186, 226

V

Valdai International Club 195
VVER-1000 reactor 19, 118, 151, 175, 176

Z

Zirconium 68, 173, 184
Zirconium-91 184
Zirconium-92 184

The manufacturer's authorised representative in the EU is Springer Nature Customer Service Centre GmbH, Europaplatz 3, 69115 Heidelberg, Germany. If you have any concerns regarding our products, please contact ProductSafety@springernature.com

Printed and bound by CPI Group (UK) Ltd, Croydon, CR0 4YY
30/09/2026
02225656-0011